Methods in Plant Molecular Biology

Mary A. Schuler and
Raymond E. Zielinski

Department of Plant Biology
University of Illinois at Urbana-Champaign
Urbana, Illinois

Academic Press, Inc.
Harcourt Brace Jovanovich, Publishers
San Diego New York Berkeley Boston
London Sydney Tokyo Toronto

ACADEMIC PRESS, INC.
San Diego, California 92101

United Kingdom Edition published by
ACADEMIC PRESS, INC. (LONDON) LTD.
24-28 Oval Road, London NW1 7DX

Library of Congress Cataloging-in-Publication Data

Schuler, Mary A.
Methods in plant molecular biology / by Mary A. Schuler, Raymond E. Zielinski.
p. cm.
Includes index.
ISBN 0-12-632340-2 (paperback) (alk. paper)
1. Plant molecular biology—Experiments. I. Zielinski, Raymond E. II. Title.
QK728.S38 1989
581.8′0724—dc19 88-12128
CIP

PRINTED IN THE UNITED STATES OF AMERICA
89 90 91 9 8 7 6 5 4 3 2

Contents

II Gel Electrophoresis Equipment

Preface

Perhaps one of the most exciting areas of modern science is the application of molecular biology to the study of plant systems. To the uninitiated scientist, trained in the classical areas of recombinant DNA technology, plant molecular biology often appears on the surface to be similar to other exciting research endeavors that explore gene structure, function, and regulation. On closer examination, however, one is impressed with the diversity of molecular techniques needed to study plant biochemistry, development, and physiology. Plant molecular biology is not simply a regurgitation of the tried and true procedures and methods used successfully for animal and bacterial cells, but requires expertise in handling organisms that have evolved formidable defenses against intrusion by an army of endogenous nucleases and proteases. This manual of laboratory methods and procedures, together with the referenced primary publications, is intended to serve both the established molecular biologist, who is attracted to exciting scientific questions in plant development and biochemistry, and those with training in classical plant physiology, who wish to utilize the powerful techniques of recombinant DNA to probe the mysteries of the plant kingdom.

This manual is an outgrowth of a semester course for advanced undergraduate and graduate students we taught at the University of Illinois at Urbana-Champaign. In this course, we have integrated many different techniques into a comprehensive format that helps students understand the diversity of molecular techniques available. We have tried to include a broader scope of protein, RNA, and DNA protocols than are currently available in recombinant DNA manuals because we feel that skilled manipulations of all types of macromolecules are essential in tackling physiological and biochemical problems. We have tried to present experiments that lead students through the technical manipulations into the fundamental, scientific questions addressed by each of these techniques. This approach strives to introduce students to chloroplast DNA structure via genomic DNA Southern analysis (Experiment 4) or to the differences in chloroplast and cytoplasmic protein synthesis via Experiments 3 and 5.

To facilitate instruction of this course, we have incorporated detailed notes for students and instructors throughout the text. Because some of the schedules for these experiments may be hard to conceive, we have included schedules outlining individual procedures to be finished in each lab segment. These schedules are especially helpful because they provide the students with definite goals for each lab period and a precise schedule for the entire semester. They also enable faculty with fewer lab periods at their disposal to pick and choose experiments tailored to their own needs. In Appendix II, we have included blueprints for gel rigs needed throughout this course.

For those attempting to unravel the mysteries of plant physiology and development, we hope these techniques facilitate the molecular dissection of plant regulatory mechanisms.

The development of this course would have been impossible without the concerted efforts of many others. We would especially like to thank Drs. Buddy Orozco and Tom Jacobs for integrating their own expertises into this course and persevering throughout its development. We also thank Drs. Fakhri Bazzaz and Tom Phillips for giving us the freedom to develop this course. We especially thank all those teaching assistants who persevered and made this course work in those first few critical years. Without the skills of Sheila Hunt, who patiently typed the manuscript innumerable times, this manual would not have materialized. Finally, we wish to thank our spouses, Stephen Sligar and Ann Zielinski, for their support and encouragement through all stages of this book.

Laboratory Schedule

In the following tabulation an outline is given that we use in the plant molecular biology laboratory course at the University of Illinois. It is based on a 15-week semester schedule with two, four-hour laboratory periods and a one-hour discussion section per week. Some experiments, however, require extra laboratory time (particularly Experiment 2). For these experiments, we schedule an additional one or two meetings per week—usually at the student's convenience—and we try to keep the necessary operations to a minimum (usually an hour or so). In some cases, if the additional manipulations are trivial (changing wash solutions, developing X-ray films, etc.), we or the teaching assistants (TAs) perform the operations for the students.

Several of the exercises in this manual have also been integrated into a plant physiology laboratory course in order to supplement the more traditional areas covered in such a course. These include Experiments 3, 4, and 7, which focus on chloroplast physiology/molecular biology. Other combinations of experiments in this manual can be used to focus on a narrower range of topics. Some possible examples are Experiments 6, 1, 2, and 7, which lead a class through the operations necessary to characterize a genomic DNA clone; Experiments 3 and 5, which illustrate the differences between the protein-synthesizing systems of the cytoplasm and chloroplasts; Experiments 1, 2, and 7, which constitute a mini-course on basic molecular cloning.

Week	Day 1	Day 2	Additional laboratory days required Day 3	Day 4
1	Experiment 1 (this experiment can be completed in one afternoon)	Experiment 2 (cut and ligate DNA)	Experiment 2 (transform *E. coli*)	
2	Experiment 2 (pick colonies; nick translate probe)	Experiment 2 (lyse colonies; bake filters; begin prehybridization)	Experiment 2 (add probe to filter hybridizations)	Experiment 2 (wash filters and start autoradiography)
3	Experiment 2 (pick colonies; start miniprep cultures) Experiment 8 (start *Agrobacterium* infection of leaf discs)	Experiment 2 (isolate DNA from minipreps; restriction cuts; run gels)		
4	Experiment 3 (prepare chloroplasts; assay chlorophyll and protein)	Experiment 3 (run SDS gels; stain and photograph gels)		
5	Experiment 3 (label plastid proteins *in vivo*)	Experiment 3 (run SDS gels; dry gels; start fluorography)		
6	Experiment 4 (perform restriction digests on chloroplast DNA isolated by TAs; start DNA isolation)	Experiment 4 (run gel of restriction fragments; photograph DNA gels; start Southerns; harvest DNA from CsCl)	Experiment 4 (bake and store Southerns)	
7	Experiment 4 (prehybridize Southerns; cut student-isolated DNA)	Experiment 4 (begin hybridization; run restriction fragments of student-isolated DNA on gels; stain)	Experiment 4 (wash Southerns and put on film)	
8	Experiment 5 (isolate RNA to step 18)	Experiment 5 (finish RNA isolation; prepare wheat germ extract)		
9	Experiment 5 (*in vitro* translation; check incorporation with TCA assay)	Experiment 5 (run SDS gels; dry gels; start fluorography)		
10	Experiment 6 (plate phage; inoculate liquid culture)	Experiment 6 (make replica filters; start prehybridization; prepare phage from liquid culture)	Experiment 6 (add probe to plaque filters)	
11	Experiment 6 (wash filters; restriction cut phage DNA; run gel)	Experiment 7 (prepare M13 phage stocks)		
12	Experiment 7 (run gels of M13 DNA; make ssDNA for sequencing reactions)	Experiment 7 (perform sequencing reactions; pour sequencing gels)	Experiment 7 (run sequencing gels; start autoradiography)	Experiment 7 (develop autoradiographs)

Week	Day 1	Day 2	Additional laboratory days required Day 3	Day 4
13	Experiment 8 (isolate DNA from *Agrobacterium*-infected calli; transfer suspension cultures for protoplasts)	Experiment 8 (restriction cut DNA from transformed calli; pour gels for genomic Southerns)		
14	Experiment 8 (run gel; set up Southerns; TAs bake and hybridize filters)	Experiment 8 (wash filters; make protoplasts)		
15	Experiment 8 (inspect autoradiograms and check regeneration in protoplasts)	Lab clean-up		

EXPERIMENT

1 Restriction Mapping of Plasmid DNA

Introduction

Restriction endonucleases are enzymes that cut DNA into discrete fragments by cleaving only at specific DNA sequences. The site at which an enzyme cuts is its "recognition sequence." Recognition sequences are generally 4, 5, or 6 base pairs (bp) and are palindromic (i.e., the sequence is the same on both strands, reading in opposite directions). The recognition sequence for the enzyme *Eco*RI is GAATTC. The palindromic nature of the sequence is seen by writing down the sequence of the double-stranded DNA at the cut site:

```
          ↓
EcoRI  5′-GAATTC-3′
       3′-CTTAAG-5′
              ↑
```

The arrows indicate the point at which the enzyme cuts the sugar–phosphate backbone of the two DNA strands. Note that *Eco*RI makes a "staggered cut" in the DNA, leaving four unpaired bases at each end of the DNA fragment (with 5′ protruding ends). Other restriction enzymes cut in the exact center of the recognition sequence leaving blunt ends:

```
            ↓
SmaI  5′-CCCGGG-3′
      3′-GGGCCC-5′
            ↑
```

Still other restriction enzymes cut the DNA so that the staggered cuts produce 3′ protruding ends:

```
              ↓
PstI  5′-CTGCAG-3′
      3′-GACGTC-5′
          ↑
```

Restriction enzymes which recognize the same sequences but

cleave at different sites within these sequences are "isoschizomers":

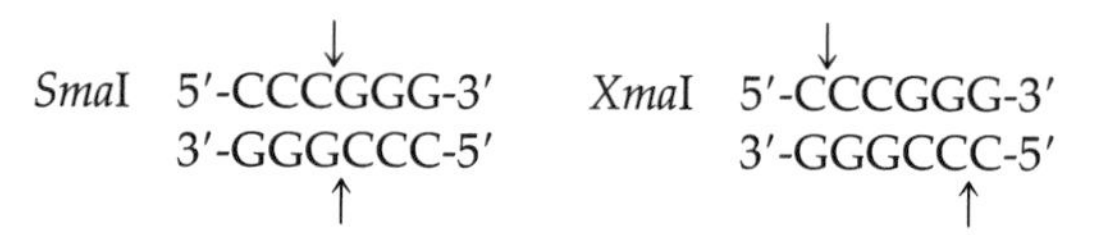

Figure 1.1 (at the end of this experiment) is taken from the *New England BioLabs* catalog. All enzymes listed in a vertical column have the same four nucleotides at the center of their recognition sequence. The column at the left designates the flanking nucleotides and the restriction cut site within this sequence. All of the enzymes in one vertical column between the heavy bars (Box 1) represent isoschizomers of one another (e.g., recognize same sequence but cut at different sites within this sequence). All of the enzymes within the same small boxes (Box 2) recognize and cut within the same sequence. All of the enzymes which occupy the same position within a larger set of boxes (two boxes marked Box 3) cleave different sequences to generate the same "sticky ends" which are "compatible" with one another in that they can hybridize with one another in DNA ligation reactions.

If the sequence of bases in DNA were random, the occurrence of a recognition sequence for any given six-base restriction enzyme would be $\frac{1}{4} \times \frac{1}{4} \times \frac{1}{4} \times \frac{1}{4} \times \frac{1}{4} \times \frac{1}{4} = 0.000244$ or once per 4096 bases. An average *Eco*RI fragment should thus be about 4000 bp or 2.5×10^6 molecular weight, if DNA sequence were random. DNA sequences are not random, of course, but it is clear from these calculations that six-base recognition enzymes are expected to make several cuts in lambda (λ) bacteriophage DNA (genome size 45×10^3 bp), several hundred cuts in *Escherichia coli* bacterial DNA (genome size 4×10^6 bp), several hundred thousand cuts in rabbit DNA (genome size 3×10^9 bp), and even more cuts in plant DNA (average genome size 10^{10} bp).

We will analyze digests of λ bacteriophage DNA and several plasmid DNAs by electrophoresis in horizontal agarose gels. A standard for calculation of molecular weights of large DNA fragments is provided by the λ bacteriophage DNA digests, whose sizes have been determined very accurately by independent methods (Figures 1.2, 1.4). There is always a very high molecular weight DNA fragment (28 kb) in the λ DNA digest which arises from the two end fragments of the λ DNA associating through their "sticky ends." If the digest is heated before electrophoresis (65°C), the fusion fragment is not seen. The molecular weights of the individual λ DNA fragments from a *Hin*dIII digestion are

*Hin*dIII	λ	23.0 kb	
		9.8 kb	
		6.6 kb	
		4.5 kb	
		2.26 kb	
		1.96 kb	
		0.53 kb	(sometimes not seen if small amounts of λ DNA are run)

The standard used for calculation of small molecular weight DNA fragments is provided by a *Hin*fI restriction digest of pBR322 DNA (Figures 1.3, 1.4).

*Hin*fI	pBR322	1630 bp	
		517 bp	
		506 bp	
		396 bp	
		344 bp	
		298 bp	
		221 bp	(two fragments)
		154 bp	
		75 bp	

The mobility of fragments in agarose gels is proportional to the logarithm of their molecular weight, at least up to a molecular weight of 10–20 kb, above which the proportionality breaks down. You should prepare, for a given agarose gel photo, a graph of the distance traveled by each λ phage DNA fragment plotted against the logarithm of its molecular weight. From this semi-log calibration curve (the lower part of which should be linear), you can infer the molecular weights of unknown DNA fragments from their mobility.

Horizontal agarose gels are usually prepared with 0.7 to 1.4% agarose. If the DNA fragments in which you are interested are rather small (less than 2 kb), a higher percentage of agarose is more desirable because the bands will be sharper and because the smaller molecular weight fragments will spread out better on the high agarose gels. On the other hand, 1.4% agarose is not useful for separating restriction digestion fragments greater than 10 kb. On this sort of gel all of the high molecular weight fragments will run together near the top of the gel. If you are working with an unknown DNA sample for the first time, it is generally advisable to use 0.8% agarose gels.

The restriction endonucleases with which you will work with at first are the enzymes *Eco*RI, *Hin*dIII, and *Bam*HI, which have six-base recognition sequences. Because enzymes can be denatured by adverse conditions, all restriction enzyme stocks should be handled carefully. Detergent, high temperature, repeated freezing and thawing, acid or alkaline pH, oxygen, vortexing, or violent shaking will all denature and inactivate the enzymes. Because of this, restriction enzymes are stored in 20–50% glycerol in −20°C freezers. They should be removed from the freezer to an ice bath just before use and the unused portion immediately returned to the −20°C freezer. Because each enzyme sample costs $50–$100 to replace, the enzymes should be pipetted by placing the end of a yellow pipet tip at the top edge of the enzyme stock solution. Dipping pipet tips down into the enzyme solutions results in inaccurate pipetting because the enzyme solution coats the outside of the pipet tips. Loss of these enzymes through carelessness will not only be expensive but extremely inconvenient for the class. **Please take care of the enzymes!!** Always keep them in an ice bath and put them back in the freezer as soon as you are finished with them!

Optimal reaction conditions for each of the enzymes are different. Each has its own preferred salt, pH, and Mg^{2+} optima. Multiple digests with two or three enzymes can be run simultaneously or serially, depending on the required conditions.

As can be seen from Figure 1.5, the three restriction enzymes used in Experiment 1 require different salt concentrations for optimal activity. Most restriction enzymes that you will be using in the class will require either the low salt (*Hin*dIII buffer), high salt (*Bam*HI buffer), or a high Tris:low salt buffer (*Eco*RI buffer), and so you will be provided with 10× stock solutions of each of these buffers to use throughout the semester. These buffers should be kept frozen when not being used (to minimize contamination and breakdown of the 2-mercaptoethanol in the buffers).

You will be provided with λ bacteriophage DNA, pBR322 DNA, pUC 13 vector DNA, and a pUC derivative that has an *Eco*RI · *Bam*HI fragment cloned into the *Eco*RI and *Bam*HI sites of the pUC linker region. You are responsible for restriction mapping the unknown DNA in the pUC vector by performing single and double restriction enzyme digests and drawing a map of the unknown plasmid DNA. Analysis of the restriction enzyme digestions will be covered in class.

Protocols

A. Restriction Digestions

1. Pour 0.8% agarose gel as follows:

a. dissolve 1.2 g agarose in 150 ml 1× agarose gel buffer (TBE)

b. heat in microwave or on hot plate until completely dissolved; cool to 50°C

c. put tape on ends of gel plate; pour agarose onto gel plate and position comb 2 inches from one edge of gel and allow to set for at least ½ hour before using*

*Student Note ▫ Throughout the course it will help if you read protocols ahead of schedule and plan on pouring agarose gels during the previous lab period. If gel is not going to be used until the next day, take the comb out and pour 1× agarose gel buffer over the gel and cover it with Saran wrap.

2. Set up the 50 μl restriction digestions outlined below, using a typical protocol such as

5 μl 10× restriction enzyme buffer
3 μl 0.3 μg/μl DNA (1 μg DNA/reaction)
41 μl sterile water
1 μl 2 U/μl restriction enzyme (1 U of enzyme cuts 1 μg DNA/hour at 37°C; this activity is measured on linear DNA, and supercoiled plasmids need a little more enzyme)
50 μl

Begin to set these digestions up by making a table that shows how much buffer, DNA, water, etc., you have to add to each tube. Check off items on your list as you add them. Never dip a previously used pipet tip into your restriction enzyme buffer or DNA stocks.

The restriction digestions that you need to set up are

1. λ DNA-*Hin*dIII
2. pBR322 DNA-*Hin*fI (use *Hin*dIII buffer)
3. unknown DNA *Eco*RI
4. unknown DNA *Bam*HI
5. unknown DNA *Hin*dIII
6. unknown DNA *Hin*dIII · *Eco*RI
7. unknown DNA *Eco*RI · *Bam*HI
8. unknown DNA *Hin*dIII · *Bam*HI
9. pUC 13 EcoRI
10. pUC 13 *Eco*RI · *Hin*dIII
11. pUC 13 *Eco*RI · *Bam*HI

(6–11) for these double restriction digestions see *Student Note

3. After setting up restriction digestions and adding the first enzyme, digest DNA at 37°C for 45 minutes. If a double enzyme restriction digest is being done, adjust NaCl or Tris concentrations as shown below, add second enzyme, and continue digestion at 37°C for 45 minutes.

*Student Note ▫ For double enzyme restriction digestions, set up first restriction digestions for enzyme with lowest salt concentration (underlined), add enzyme, and incubate at 37°C for 45 minutes. Then adjust NaCl or Tris concentrations as follows after first reaction is completed:

10× *Hin*dIII buffer	10× *Eco*RI buffer	10× *Bam*HI buffer
*Hin*dIII	*Eco*RI	*Bam*HI
*Pst*I	*Ava*I	*Sal*I
*Bgl*II		
*Hin*fI		
*Bgl*I		

10× *Eco*RI buffer ⟶ to adjust for *Bam*HI digestion, add 2.5 μl 2 M NaCl/50 μl

10× *Hin*dIII buffer ⟶ to adjust for *Eco*RI digestion, add 5 μl 1 M Tris (pH 7.4)/50 μl

10× *Hin*dIII buffer ⟶ to adjust for *Bam*HI digestion, add 2.5 μl 2 M NaCl and 5 μl 1 M Tris (pH 7.4)/50 μl

4. At the end of these incubations, add 10 μl 10 *M* urea loading dye to each sample, mix, and load 15 μl of each sample per lane on gel (0.15 μg DNA/lane).
5. After loading samples, electrophorese gel for 2 hours at 65 mA/gel. If several gels are run on the same power supply, double or triple the current depending on the number of gels hooked in series. Don't forget to load λ *Hin*dIII cut and pBR322 *Hin*fI cut standards.
6. After the gel is finished running, take gel off horizontal plate and put in Pyrex baking dish; cover with water; add 10 μl 1 mg/ml ethidium bromide (EtBr) to gel. **Ethidium bromide is a carcinogen, so wear gloves at all times when handling it! Also, chemically inactivate EtBr solutions before pouring them down the drain.** Allow gel to stain for 20 minutes at room temperature; photograph with Polaroid type 57 film.

B. Photographing the Gel

1. **Put on UV goggles to protect your eyes. Short wavelength UV light from the transilluminator will sunburn your corneas unless you wear goggles.**
2. Examine your gel (room lights out, UV light on). At *f*4.5, the exposure time needed will vary from 3 seconds (extremely bright bands) to 30 seconds (bands that you can hardly see). Turn UV off, rooms lights on.
3. Place a black-lined ruler on the gel; focus the camera using floodlights for accurate focus.
4. Move camera back over lens. Set the *f*/stop to 4.5 and put orange filter over lens; turn off room lights, turn on UV lights, and expose picture for appropriate time. Pull out white tab and smoothly pull out picture tab. Wait 20 seconds, then peel backing off your picture.

C. Data Interpretation

1. Construct a standard curve for the DNA fragments from the λ DNA and pBR322 DNAs. Plot log (molecular weight in kilobases or base pairs) vs R_f (distance DNA fragment moved/distance bromphenol blue moved).

2. Determine the molecular weights of the digest fragments in your unknown DNA sample. By comparison with the pUC vector digest, deduce which fragments come from the vector and which from the insert fragments.

The total molecular weight of fragments in every digest of the plasmid should be the same and should equal the size of the vector plus the insert. If any digest gives a smaller total molecular weight, examine it carefully for bright bands that might be doublets or triplets which account for the missing restriction fragments. Some small fragments may disappear from the gel if they migrate with the bromphenol blue tracking dye. These small fragments can usually be resolved on higher percentage agarose gels.

The total number of fragments in a digestion should be additive (i.e., if *Bam*HI makes one cut and *Eco*RI makes two cuts, then *Eco*RI and *Bam*HI digestions together must make three cuts and there should be three fragments on the gel). Always check to see whether you have the expected number of DNA bands in a restriction digest. If you do not have the predicted number of bands, then two DNA fragments are migrating at the same place on the gel *or* two of the restriction sites lie very close to one another *or* one of your enzymes isn't cutting the DNA.

Materials Provided

0.5 μg/μl λ DNA (standard)
0.5 μg/μl pBR322 DNA (standard)
0.3 μg/μl unknown plasmid DNA (*Eco*RI · *Bam*HI DNA fragment cloned into *Eco*RI · *Bam*HI sites on pUC 13)
Sterile water
*Hin*dIII, *Eco*RI, *Bam*HI restriction enzymes diluted to 2 U/μl
Agarose
1 mg/ml ethidium bromide in water

10 *M* urea loading dye

10 *M* urea	60 g urea
0.1% bromphenol blue	0.1 g bromphenol blue
0.1% xylene cyanol	0.1 g xylene cyanol
	up to 100 ml with sterile water

Solution	Preparation
10× agarose gel buffer (TBE)	108 g Tris base
	9.3 g Na_2EDTA
	55 g boric acid
	up to 1 liter with distilled water
10× *Hin*dIII buffer	
500 m*M* NaCl	2.9 g NaCl
60 m*M* Tris–HCl (pH 7.4)	6 ml 1 *M* Tris–HCl (pH 7.4)
60 m*M* $MgCl_2$	1.2 g $MgCl_2 \cdot 6H_2O$
60 m*M* 2-mercaptoethanol	430 μl 2-mercaptoethanol
1 mg/ml BSA	0.1 g BSA
	up to 100 ml with sterile water
10× *Eco*RI buffer	
1000 m*M* Tris–HCl (pH 7.4)	12.1 g Tris base
500 m*M* NaCl	2.9 g NaCl
50 m*M* $MgCl_2$	1.0 g $MgCl_2 \cdot 6H_2O$
1 mg/ml BSA	0.1 g BSA
	add up to 95 ml with sterile water
	add conc. HCl until pH is 7.4 (5 ml or more)
	up to 100 ml with sterile water
10× *Bam*HI buffer	
1500 m*M* NaCl	8.8 g NaCl
60 m*M* Tris–HCl (pH 7.9)	6 ml 1 *M* Tris–HCl (pH 7.9)
60 m*M* $MgCl_2$	1.2 g $MgCl_2 \cdot 6H_2O$
1 mg/ml BSA	0.1 g BSA
	up to 100 ml with sterile water
2 *M* NaCl	11.7 g NaCl
	up to 100 ml with sterile water

References

Freifelder, D. (1983). Restriction endonucleases. *In* "Molecular Biology" (K. Sergent, ed.), 2nd ed., pp. 124–129. Jones and Bartlett, Portola Valley, California.

New England BioLabs Catalog 1986/87. Beverly, Massachusetts. 135 pp.

Rodriquez, R. L., and Tait, R. C. (1983). Restriction endonucleases. "Recombinant DNA Techniques: An Introduction," pp. 53–66. Addison-Wesley, Reading, Massachusetts.

Rodriquez, R. L., and Tait, R. C. (1983). Gel electrophoresis. "Recombinant DNA Techniques: An Introduction," pp. 67–79. Addison-Wesley, Reading, Massachusetts.

Cross Index of Palindromic and Related Recognition Sequences

Sequences at the top of each column are written 5′ to 3′ according to convention. Open squares at the left of each row are place holders for nucleotides within a restriction endonuclease recognition sequence, and the arrowheads indicate the point of cleavage. Sequences of complementary strands and their cleavage sites are implied. Enzymes written in red bold type recognize only one sequence. Enzymes written in light type have multiple recognition sequences. These enzymes with all their recognition sequences and cleavage sites are listed on the opposite page. An asterisk (*) indicates a sequence cleaved identically by two or more enzymes that are affected differently by DNA modification at that site (Roberts, 1982), e.g., *Hpa* II and *Msp* I recognize and cleave at the same site but are affected differently by DNA modification. *Mbo* I and *Sau*3A I share a similar relationship.

Palindromic Tetra- and Hexa-Nucleotide Recognition Sequences and Cleavage Sites for Restriction Endonucleases

	AATT	ACGT	AGCT	ATAT	CATG	CCGG	CGCG	CTAG	GATC	GCGC	GGCC	GTAC	TATA	TCGA	TGCA	TTAA
▼□□□□									Mbo I*							
□▼□□□		Mae II				Hpa II*		Mae I		HinP I				Taq I		Mse I
□□▼□□			Alu I				BstU I		Dpn I		Hae III	Rsa I				
□□□▼□										Hha I						
□□□□▼					Nla III				Box 3							
A▼□□□□T			Hind III		Afl III		Mlu I Afl III	Spe I	Bgl II BstY I							
A□▼□□□T														Cla I		Ase I
A□□▼□□T				Ssp I						Eco47 III	Stu I	Sca I				
A□□□▼□T																
A□□□□▼T					Nsp7524I	Box 1				Hae II					Nsi I	
C▼□□□□G					Nco I Sty I	Xma I Ava I		Avr II Sty I			Eag I Eae I Gdi II			Xho I Ava I		Afl II
C□▼□□□G				Nde I												
C□□▼□□G			Pvu II NspB II			Sma I	NspB II									
C□□□▼□G							Sac II		Pvu I							
C□□□□▼G									Box 3						Pst I	
G▼□□□□C	EcoR I						BssH II	Nhe I	BamH I BstY I	Ban I		Asp718 Ban I		Sal I	ApaL I	
G□▼□□□C		Aha II								Nar I Aha II			Acc I	Acc I		
G□□▼□□C				EcoR V		Nae I						Xca I		Hinc II		Hpa I Hinc II
G□□□▼□C										Box 2						
G□□□□▼C		Aat II	Sac I Ban II HgiA I Bsp 1286		Sph I Nsp7524 I					Bbe I Hae II	Apa I Ban II Bsp1286	Kpn I			Bsp1286 HgiA I	
T▼□□□□A					BspH I	BspM II		Xba I	Bcl I		Eae I					
T□▼□□□A														BstB I		
T□□▼□□A		SnaB I					Nru I			Fsp I	Bal I					Dra I
T□□□▼□A																
T□□□□▼A																

Figure 1.1. Cross index of palindromic and related recognition sequences. (Courtesy of New England BioLabs, Inc.)

Bacteriophage Lambda Restriction Map

Lambda is a large *E. coli* bacteriophage. The DNA molecule in the virion is linear and, except for the extreme left and right ends, double stranded. At each end the 5′ strand overhangs the 3′ strand by 12 bases giving rise to a short single stranded terminus. The nucleotide sequences of the single stranded ends are complementary; they anneal rapidly in vivo and in vitro leading to the formation of circular, completely double stranded molecules. The circular form of the molecule is 48502 base pairs in length. By convention, numbering of the nucleotide sequence begins with the first base of the left cohesive end: GGGCGGCGACCT . . . and increments 5′ to 3′ along the 'L' strand in the direction late genes to early genes. Numbering of the nucleotide sequence stops at nucleotide 48502, the 3′ end of the 'L' strand; it does not include the right cohesive end which extends 12 bases further on the complementary strand. The table at right lists the locations of some of the restriction sites in wild type lambda DNA. Data have only been included for those enzymes which cut the DNA a modest number of times. The table entries give the coordinate of the base which corresponds to the 5′ nucleotide of each recognition sequence.

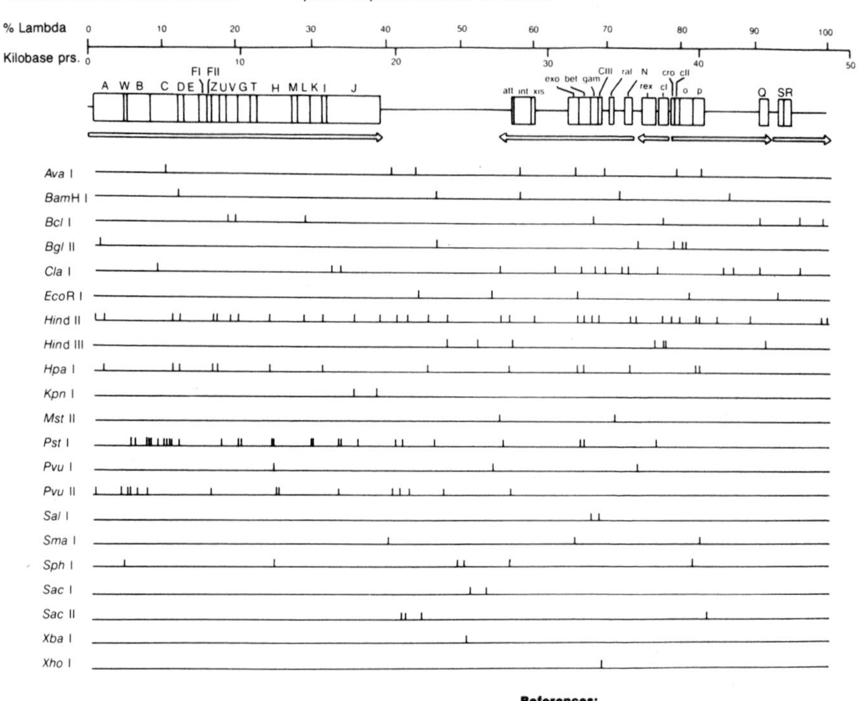

References:

1. Echols, H. and Murialdo, H., (1978) *Microbiol. Rev. 42*, 577–591
2. Szybalski, E.H. and Szybalski, W., (1979) *Gene 7*, 217–270
3. Daniels, D.L. de Wet, J.R. and Blattner, F.R., (1980) *J. Virol. 33*, 390–400
4. Sanger, F., Coulson, A.R., Hong, G.F., Hill, D.F. and Petersen, G.B., (1982) *J. Mol. Biol. 162*, 729–773
5. Daniels, D.L., Schroeder, J.L., Blattner, F.R., Szybalski, W. and Sanger, F., (1983) in: Hendrix, R.W., Roberts, J.W., Stahl, F.W. and Weisberg, R.A. (eds) *Lambda II:* Appendix; Cold Springs Harbor, New York
6. Roberts, R.J., (1983) *Nucleic Acids Res. 11*, r135–r167
7. *GENBANK 7.0*, (April 1, 1983)

Figure 1.2. Bacteriophage λ restriction map. (Courtesy of New England BioLabs, Inc.)

LAMBDA DNA — Location of Restriction Sites

Enzyme	#	Locations				
Apa I	1	10086				
Nae I	1	20040				
Nar I	1	45679				
Xba I	1	24508				
Xho I	1	33498				
Avr II	2	24322	24396			
Kpn I	2	17053	18556			
Mst II	2	26717	34318			
Sac I	2	24772	25877			
Sal I	2	32745	33244			
Tth111 I	2	11202	36120			
Xma III	2	19944	36654			
Afl II	3	6540	12618	42630		
Pvu I	3	11933	26254	35787		
Sma I	3	19397	31617	39888		
Sna I	3	15260	18834	19473		
Nco I	4	19329	23901	27868	44248	
Sac II	4	20320	20530	21606	40386	
BamH I	5	5505	22346	27972	34499	41732
EcoR I	5	21226	26104	31747	39168	44972
Nru I	5	4590	28050	31703	32407	41808
Sca I	5	16421	18684	25685	27263	32802
Bgl II	6	415	22425	35711	38103	38754
		38814				
BssH II	6	3522	4126	5627	14815	16649
		28008				
Hind III	6	23130	25157	27479	36895	37459
		44141				
Sph I	6	2212	12002	23942	24371	27374
		39418				
Stu I	6	12434	31478	32997	39992	40596
		40614				
Asu II	7	18048	25884	27980	29150	30396
		34331	42637			
Ban II	7	581	10086	19763	21570	24772
		25877	39453			
Mlu I	7	458	5548	15372	17791	19996
		20952	22220			
Nde I	7	27630	29883	33679	36112	36668
		38357	40131			
Ava I	8	4720	19397	20999	27887	31617
		33498	38214	39888		
Bcl I	8	8844	9361	13820	32729	37352
		43682	46366	47942		
Acc I	9	2190	15260	18834	19473	31301
		32745	33244	40201	42921	
Aat II	10	5105	9394	11243	14974	29036
		40806	41113	42247	45563	45592
Aha III	13	90	8460	16294	23110	23284
		25436	26132	26665	32703	36302
		36530	38833	47429		
BstE II	13	5687	7058	8322	9024	13348
		13572	13689	16012	17941	25183
		30005	36374	40049		
BstX I	13	2855	6706	8413	8850	10915
		13263	14338	18029	19741	21622
		34596	38292	46434		
Ava III	14	10325	27206	27372	28432	30342
		30989	32967	33682	34208	35868
		36665	36671	37769	38307	
HgiE II	14	1785	2250	5903	6555	12513
		13954	15877	17433	20244	26435
		35595	35639	37999	42048	
Hpa I	14	732	5267	5708	7948	8199
		11583	14991	21902	27316	31807
		32217	35259	39606	39834	
Cla I	15	4198	15583	16120	26616	30289
		31990	32963	33584	34696	35050
		36965	41363	42020	43824	46438
Mst I	15	463	2503	4270	5155	6979
		11563	11690	13355	16046	21805
		21826	27949	32683	34821	42380
Pvu II	15	209	1917	2385	2526	3058
		3637	7831	12099	12162	16078
		19716	20059	20695	22991	27412
Bal I	18	1326	2206	3260	4193	6496
		6877	7584	7978	8056	8859
		10609	10777	13934	14903	21260
		26623	28618	36040		
Afl III	20	458	628	5548	11281	15372
		17791	18284	19996	20952	22220
		24133	24168	26528	32764	39395
		42086	42363	43762	44501	46982
EcoR V	21	650	2084	6681	8084	8822
		13435	14023	17767	18385	21269
		22948	26821	28198	28211	33587
		39352	41273	41541	41576	42231
		45826				
Gdi II	21	2739	5601	6008	8366	10588
		13481	14575	16416	18547	19284
		19332	19944	20239	20323	20928
		20988	22025	35465	36654	39458
		45214				
Xho II	21	415	1606	2531	5505	6422
		22346	22425	24511	27027	27972
		29593	30426	34499	35711	38103
		38664	38754	38814	39576	41732
		47773				
Xmn I	24	33	1151	2319	8490	10111
		13102	16909	22852	22871	23808
		23828	24228	24578	25485	27252
		29015	29993	31085	33811	34185
		42477	44727	45741	47564	
Ban I	25	1180	1365	2331	5407	5665
		5671	5900	8036	8043	8441
		8764	8988	10221	13038	13642
		14623	15199	15237	16236	17053
		18556	21545	39907	42797	45679
HgiA I	28	5619	6002	9485	10295	11950
		13289	13492	14474	15211	16516
		21612	21798	21852	24772	25877
		26469	27173	33467	35583	37933
		40216	40489	42371	42512	44177
		44846	46698	47660		
Pst I	28	2556	2820	3625	3640	3856
		4370	4709	4909	5120	5214
		5682	8520	9613	9777	11763
		11835	14294	14381	16081	16231
		17390	19833	20281	22421	26928
		32005	32252	37001		
Bgl I	29	404	2660	3798	4360	4451
		4577	5246	5432	6053	6104
		7550	8049	11058	12708	12717
		12832	13198	14401	14890	15157
		17638	18085	19334	20124	20250
		20460	21233	30882	32323	
NspC I	32	628	2212	6478	8375	12002
		17274	18758	21802	23425	23942
		24371	25099	25659	25868	27374
		29170	30738	31542	32493	33717
		34649	35082	38021	38996	39395
		39418	39646	40069	42346	43068
		46123	47840			
Ava II	35	1612	1922	2816	3801	4314
		4622	6042	6440	8995	11000
		11045	12996	13147	13737	13952
		13984	14329	15613	16587	16610
		16683	19289	19356	19867	22001
		22243	28798	32474	32562	39004
		39437	39479	47605	48202	48474
Hinc II	35	197	732	5267	5708	7948
		8199	9054	9624	11583	13783
		14991	17074	18754	19839	20567
		21902	23145	26742	27316	28926
		31807	32217	32745	33244	35259
		35613	37431	37987	38546	39606
		39834	40940	43181	47936	48296
Bsp1286	38	581	5619	5664	6002	9485
		10086	10295	11414	11950	13039
		13289	13492	14474	14897	15211
		16516	19763	21570	21612	21798
		21852	24772	25877	26469	27173
		32330	33467	35583	37933	39453
		40216	40489	42371	42512	44177
		44846	46698	47660		
Cfr I	39	1326	2206	2739	3260	4193
		5601	6008	6496	6877	7584
		7978	8056	8366	8859	10588
		10609	10777	13481	13934	14575
		14903	16416	18547	19284	19332
		19944	20239	20323	20928	20988
		21260	22025	26623	28618	35465
		36040	36654	39458	45214	
Aha II	40	1475	1496	2303	4947	4985
		5105	6915	8096	8263	9089
		9394	9452	9861	10080	10621
		11243	11768	12929	13318	14799
		14974	16056	17616	17670	28467
		29036	30472	30727	31765	31936
		35072	40806	41113	42247	44221
		44330	44912	45563	45592	45679

Restriction Map of pBR322 DNA

4363 base pairs

pBR322 is an *E. coli* plasmid cloning vehicle. The molecule is a double-stranded DNA circle 4363 base pairs in length. pBR322 was constructed *in vitro* from the tetracycline resistance gene (Tc) from pSC101, the origin of DNA replication (Ori) from the colE1 derivative pMB1, and the ampicillin resistance gene (Ap) from transposon Tn3. Numbering of the sequence begins within the unique *Eco*R I site: the first T in the sequence . . . GAATTC . . . is designated as nucleotide number 1. Numbering then continues around the molecule in the direction Tc to Ap.

The map shows the restriction sites of those enzymes that cut the molecule once or twice; the unique sites are shown in bold type. The table lists the sites of those enzymes that cut the molecule a moderate number of times. The entries refer to the coordinate of the first (5′) base in the recognition sequence. The map also shows the positions of the antibiotic resistance genes and the origin of replication.

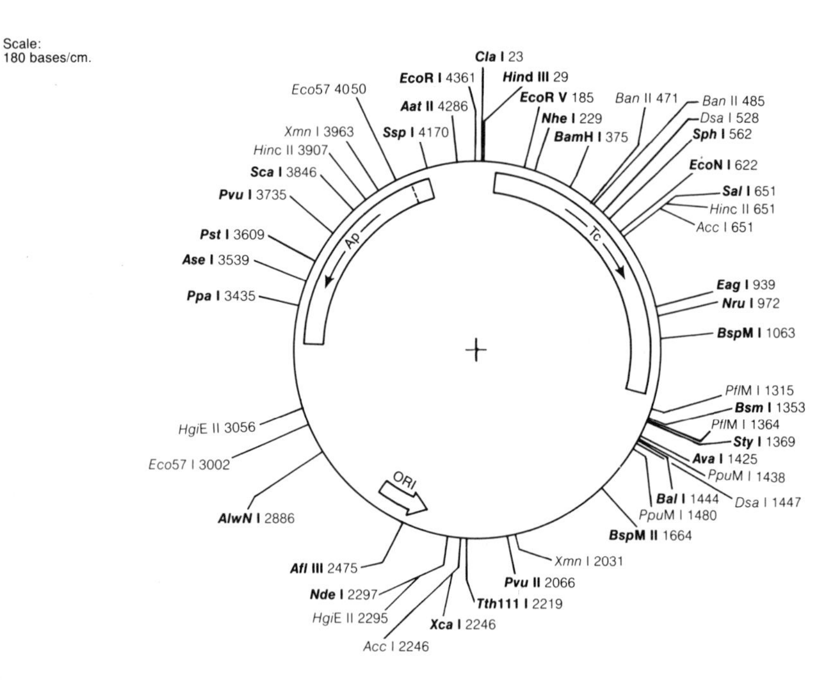

Reference
1. Bolivar, F., Rodriguez, R.L., Greene, P.J., Betlach, M.C., Heynecker, H.L. and Boyer, H.W. (1977) *Gene* 2, 95–113
2. Sutcliffe, J.G. (1978) *Cold Spring Harb. Symp. Quant. Biol.* 43, 77–90
3. Sutcliffe, J.G. (1978) *Proc. Natl. Acad. Sci.* USA 75, 3737–3741
4. Peden, K.W.C. (1983) *Gene* 22, 277–280
5. Backman, K. and Boyer, H.W. (1983) *Gene* 26, 197–203
6. Lathe, R., Kieny, M.P., Skory, S. and Lecocoq, J.P. (1984) *DNA* 3, 173–182
7. Heusterspreute, M. and Davison, J. (1984) *DNA* 3, 259–264
8. Roberts, R.J. (1987) *Nucleic Acids Res.* 15, Supplement, r189–r217
9. GenBank (1987) 50.0

Figure 1.3. Restriction map of pBR322 DNA. (Courtesy of New England BioLabs, Inc.)

pBR322 DNA–Location of Restriction Sites

Enzyme	#	Locations				
Aat II	1	4286				
Afl III	1	2475				
AlwN I	1	2886				
Ase I	1	3539				
Ava I	1	1425				
Bal I	1	1444				
BamH I	1	375				
Bsm I	1	1353				
BspM I	1	1063				
BspM II	1	1664				
Cla I	1	23				
Eag I	1	939				
EcoN I	1	622				
EcoR I	1	4361				
EcoR V	1	185				
Hind III	1	29				
Nde I	1	2297				
Nhe I	1	229				
Nru I	1	972				
Ppa I	1	3435				
Pst I	1	3609				
Pvu I	1	3735				
Pvu II	1	2066				
Sal I	1	651				
Sca I	1	3846				
Sph I	1	562				
Ssp I	1	4170				
Sty I	1	1369				
Tth111 I	1	2219				
Xca I	1	2246				
Acc I	2	651	2246			
Ban II	2	471	485			
Dsa I	2	528	1447			
Eco57 I	2	3002	4050			
HgiE II	2	2295	3056			
Hinc II	2	651	3907			
PflM I	2	1315	1364			
PpuM I	2	1438	1480			
Xmn I	2	2031	3963			
ApaL I	3	2291	2789	4035		
Bbv II	3	737	1600	4353		
Bgl I	3	929	1163	3482		
Dra I	3	3232	3251	3943		
Rsa I	3	164	2282	3847		
Ait I	4	232	494	775	1727	
BspH I	4	489	3195	4203	4308	
Eco0109 I	4	523	1438	1480	4343	
Fin I	4	538	888	1084	1761	
Fsp I	4	260	1356	1454	3588	
Gsu I	4	811	1401	1983	3453	
Mme I	4	197	284	2665	2849	
Nae I	4	401	769	929	1283	
Nar I	4	413	434	548	1205	
Nsp7524 I	4	562	1816	2110	2475	
Ple I	4	632	2375	2846	3363	
Gdi II	5	295	399	531	939	3756
Mae I	5	230	1489	2970	3223	3558
Tth111 II	5	7	1922	3049	3082	3088
Aha II	6	413	434	548	1205	3904
		4286				
BstN I	6	130	1059	1442	2502	2623
		2636				
Eae I	6	295	399	531	939	1444
		3756				
NspB II	6	1139	2066	2185	2815	3060
		4001				
Taq II	6	654	2387	3726	3885	4038
		4081				
Cfr10 I	7	160	401	410	769	929
		1283	3448			
Taq I	7	24	339	652	1127	1268
		2575	4019			
Ava II	8	799	887	1136	1439	1481
		1760	3506	3728		
BstY I	8	375	1667	3116	3127	3213
		3225	3993	4010		
Dde I	8	1581	1743	2285	2750	3159
		3325	3865	4291		
HgiA I	8	276	587	1174	1465	2291
		2789	3950	4035		
Sec I	8	115	129	528	534	1167
		1369	1447	2635		
Ban I	9	76	119	413	434	548
		766	1205	1289	3316	
Bsp1286 I	10	276	471	485	587	1174
		1465	2291	2789	3950	4035
Hinf I	10	632	852	1006	1304	1525
		2031	2375	2450	2846	3363

Enzyme	#	Locations				
Mae II	10	901	957	1546	1570	1800
		2228	3178	3594	3967	4287
Nci I	10	170	534	1258	1484	1812
		2120	2155	2854	3550	3901
Hae II	11	232	413	434	494	548
		775	1205	1644	1727	2349
		2719				
Hga I	11	390	649	944	976	1240
		1390	2004	2181	2577	3155
		3905				
Mbo II	11	464	738	1009	1601	2354
		3125	3216	3971	4049	4158
		4354				
Alw I	12	375	376	1097	1667	3042
		3116	3128	3213	3226	3690
		3993	4011			
Fok I	12	112	133	987	1032	1681
		1770	1848	2009	2150	3348
		3529	3816			
Hph I	12	126	408	453	1307	1528
		2085	2094	3219	3446	3842
		4068	4083			
Sau96 I	15	172	524	799	887	1136
		1260	1439	1481	1760	1949
		3410	3489	3506	3728	4344
ScrF I	16	130	170	534	1059	1258
		1442	1484	1812	2120	2155
		2502	2623	2636	2854	3550
		3901				
Alu I	17	15	30	686	1089	1999
		2056	2067	2116	2135	2416
		2642	2732	2778	3035	3556
		3656	3719			
Mae III	17	125	213	881	1148	1808
		1831	1917	2130	2225	2832
		2895	3011	3294	3625	3683
		3836	4024			
Bbv I	21	226	615	773	1406	1430
		1559	1562	1685	2065	2068
		2114	2211	2380	2398	2817
		2882	2885	3091	3419	3608
		3785				
Hae III	22	173	296	400	524	532
		596	830	919	940	991
		1048	1261	1445	1949	2489
		2500	2518	2952	3410	3490
		3757	4344			
Mbo I	22	349	376	467	826	1098
		1129	1144	1461	1668	3042
		3117	3128	3136	3214	3226
		3331	3672	3690	3736	3994
		4011	4047			
SfaN I	22	134	204	247	393	405
		658	1026	1033	1421	1673
		1682	1769	1847	1910	2151
		2267	2322	2343	2563	3615
		3825	4055			
BstU I	23	346	702	817	946	973
		978	1039	1105	1234	1244
		1389	1415	1537	1634	2006
		2075	2077	2180	2521	3102
		3432	3925	4257		
Nla IV	24	76	119	329	375	413
		434	524	548	766	887
		1205	1254	1289	1324	1438
		1481	1760	2505	2544	3316
		3410	3451	3662	4252	
Hpa II	26	161	170	387	402	411
		534	694	770	930	1020
		1258	1284	1485	1665	1812
		2121	2155	2682	2829	2855
		3045	3449	3483	3550	3660
		3902				

There are no restriction sites for the following enzymes in PBR322 DNA :

Afl II	Apa I	Avr II	Bcl I
Bgl II	BssH II	BstB I	BstE II
BstX I	Bsu36 I	Dra III	Esp I
Hpa I	Kpn I	Mlu I	Nco I
Not I	Nsi I	Pfu I	PmaC I
Rsr II	Sac I	Sac II	Sfi I
Sma I	SnaB I	Spe I	Stu I
Xba I	Xho I		

SIZES OF RESTRICTION FRAGMENTS OF pBR322[a,b]

*Hae*III	*Hpa*II	*Alu*I	*Hinf*I	*Taq*I	*Tha*I	*Hha*I	*Hae*II	*Mbo*I
587	622	910	1631	1444	581	393	1876	1374
540	527	659	517	1307	493	347	622	665
504	404	655	506	475	452	337	439	358
458	309	521	396	368	372	332	430	341
434	242	403	344	315	355	270	370	317
267	238	281	298	312	341	259	227	272
234	217	257	221	141	332	206	181	258
213	201	226	220		330	190	83	207
192	190	136	154		145	174	60	105
184	180	100	75		129	153	53	91
124	160	63			129	152	21	78
123	160	57			122	151		75
104	147	49			115	141		46
89	147	19			104	132		36
80	122	15			97	131		31
64	110	11			68	109		27
57	90				66	104		18
51	76				61	100		17
21	67				27	93		15
18	34				26	83		12
11	34				10	75		11
7	26				5	67		8
	26				2	62		
	15					60		
	9					53		
	9					40		
						36		
						33		
						30		
						28		
						21		

[a] Data from Sutcliffe, J. G. (1978). *Cold Spring Harbor Symp. Quant. Biol.* **43,** 77–90.

[b] These sizes (in base pairs) do not include any extension which may be left by a particular enzyme.

SIZES OF RESTRICTION FRAGMENTS OF PHAGE λ DNA[a,b]

Fragment	*Eco*RI	*Hin*dIII	*Eco*RI plus *Hin*dIII	*Bgl*II	*Ava*I
A	21.8	23.7	21.8	22.8	15.9
B	7.52	9.46	5.24	13.6	8.8
C	5.93	6.75 (6.61)	5.05	9.8	6.1
D	5.54	4.26	4.21	2.3	4.6 (two fragments)
E	4.80	2.26	3.41	0.46	4.1
F	3.41	1.98	1.98		1.8
G		0.58	1.90		1.61
H			1.71 (1.57)		1.55
I			1.32		
J			0.93		
K			0.84		
L			0.58		

[a] All values are in kilobases and the full size of λ DNA is taken as 49 kb.

[b] Data from Hendrix, R. W., Roberts, J. W., Stahl, F. W., and Weisberg, R. A., eds. (1983). "Lambda II." Cold Spring Harbor Laboratory, Cold Spring Harbor, New York.

Figure 1.4. Sizes of λ and pBR322 DNA restriction fragments.

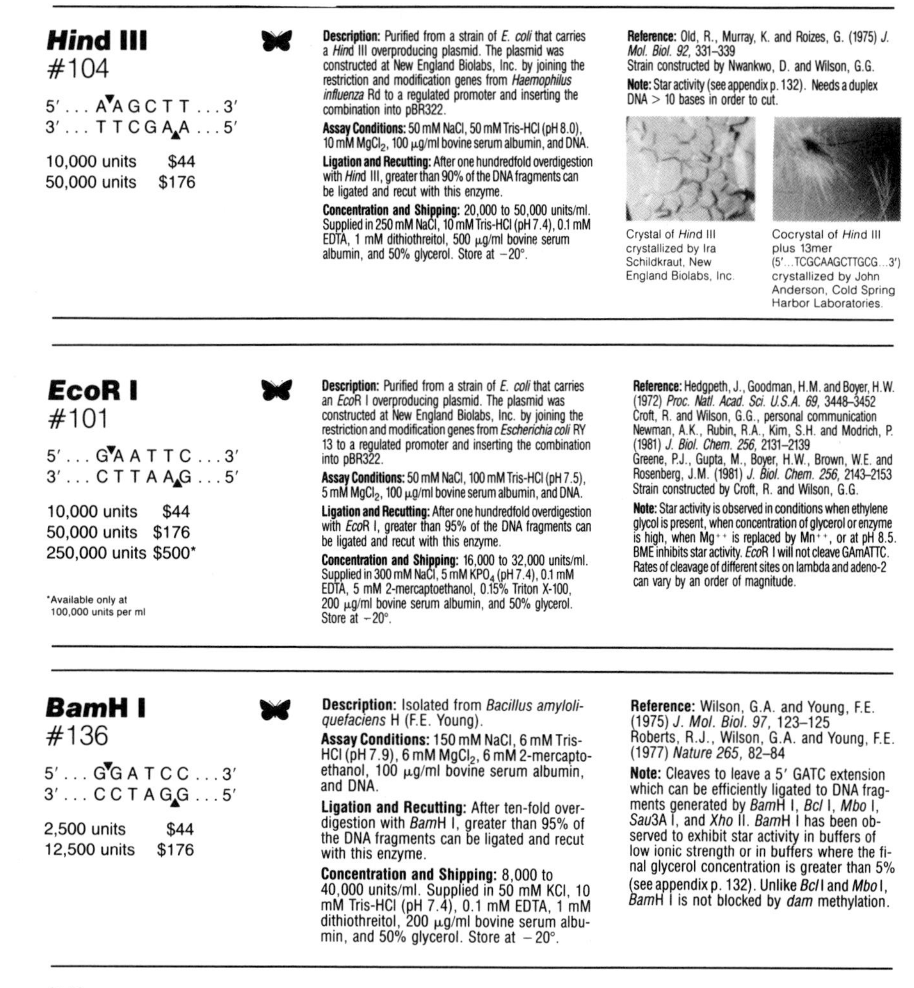

Hind III
#104

5′ . . . A▼A G C T T . . . 3′
3′ . . . T T C G A▲A . . . 5′

10,000 units $44
50,000 units $176

Description: Purified from a strain of *E. coli* that carries a *Hind* III overproducing plasmid. The plasmid was constructed at New England Biolabs, Inc. by joining the restriction and modification genes from *Haemophilus influenza* Rd to a regulated promoter and inserting the combination into pBR322.

Assay Conditions: 50 mM NaCl, 50 mM Tris-HCl (pH 8.0), 10 mM $MgCl_2$, 100 μg/ml bovine serum albumin, and DNA.

Ligation and Recutting: After one hundredfold overdigestion with *Hind* III, greater than 90% of the DNA fragments can be ligated and recut with this enzyme.

Concentration and Shipping: 20,000 to 50,000 units/ml. Supplied in 250 mM NaCl, 10 mM Tris-HCl (pH 7.4), 0.1 mM EDTA, 1 mM dithiothreitol, 500 μg/ml bovine serum albumin, and 50% glycerol. Store at −20°.

Reference: Old, R., Murray, K. and Roizes, G. (1975) *J. Mol. Biol. 92*, 331–339
Strain constructed by Nwankwo, D. and Wilson, G.G.

Note: Star activity (see appendix p. 132). Needs a duplex DNA > 10 bases in order to cut.

Crystal of *Hind* III crystallized by Ira Schildkraut, New England Biolabs, Inc.

Cocrystal of *Hind* III plus 13mer (5′...TCGCAAGCTTGCG...3′) crystallized by John Anderson, Cold Spring Harbor Laboratories.

EcoR I
#101

5′ . . . G▼A A T T C . . . 3′
3′ . . . C T T A A▲G . . . 5′

10,000 units $44
50,000 units $176
250,000 units $500*

*Available only at 100,000 units per ml

Description: Purified from a strain of *E. coli* that carries an *EcoR* I overproducing plasmid. The plasmid was constructed at New England Biolabs, Inc. by joining the restriction and modification genes from *Escherichia coli* RY 13 to a regulated promoter and inserting the combination into pBR322.

Assay Conditions: 50 mM NaCl, 100 mM Tris-HCl (pH 7.5), 5 mM $MgCl_2$, 100 μg/ml bovine serum albumin, and DNA.

Ligation and Recutting: After one hundredfold overdigestion with *EcoR* I, greater than 95% of the DNA fragments can be ligated and recut with this enzyme.

Concentration and Shipping: 16,000 to 32,000 units/ml. Supplied in 300 mM NaCl, 5 mM KPO_4 (pH 7.4), 0.1 mM EDTA, 5 mM 2-mercaptoethanol, 0.15% Triton X-100, 200 μg/ml bovine serum albumin, and 50% glycerol. Store at −20°.

Reference: Hedgpeth, J., Goodman, H.M. and Boyer, H.W. (1972) *Proc. Natl. Acad. Sci. U.S.A. 69*, 3448–3452
Croft, R. and Wilson, G.G., personal communication
Newman, A.K., Rubin, R.A., Kim, S.H. and Modrich, P. (1981) *J. Biol. Chem. 256*, 2131–2139
Greene, P.J., Gupta, M., Boyer, H.W., Brown, W.E. and Rosenberg, J.M. (1981) *J. Biol. Chem. 256*, 2143–2153
Strain constructed by Croft, R. and Wilson, G.G.

Note: Star activity is observed in conditions when ethylene glycol is present, when concentration of glycerol or enzyme is high, when Mg^{++} is replaced by Mn^{++}, or at pH 8.5. BME inhibits star activity. *EcoR* I will not cleave GAmATTC. Rates of cleavage of different sites on lambda and adeno-2 can vary by an order of magnitude.

BamH I
#136

5′ . . . G▼G A T C C . . . 3′
3′ . . . C C T A G▲G . . . 5′

2,500 units $44
12,500 units $176

Description: Isolated from *Bacillus amyloliquefaciens* H (F.E. Young).

Assay Conditions: 150 mM NaCl, 6 mM Tris-HCl (pH 7.9), 6 mM $MgCl_2$, 6 mM 2-mercaptoethanol, 100 μg/ml bovine serum albumin, and DNA.

Ligation and Recutting: After ten-fold overdigestion with *BamH* I, greater than 95% of the DNA fragments can be ligated and recut with this enzyme.

Concentration and Shipping: 8,000 to 40,000 units/ml. Supplied in 50 mM KCl, 10 mM Tris-HCl (pH 7.4), 0.1 mM EDTA, 1 mM dithiothreitol, 200 μg/ml bovine serum albumin, and 50% glycerol. Store at −20°.

Reference: Wilson, G.A. and Young, F.E. (1975) *J. Mol. Biol. 97*, 123–125
Roberts, R.J., Wilson, G.A. and Young, F.E. (1977) *Nature 265*, 82–84

Note: Cleaves to leave a 5′ GATC extension which can be efficiently ligated to DNA fragments generated by *BamH* I, *Bcl* I, *Mbo* I, *Sau3A* I, and *Xho* II. *BamH* I has been observed to exhibit star activity in buffers of low ionic strength or in buffers where the final glycerol concentration is greater than 5% (see appendix p. 132). Unlike *Bcl* I and *Mbo* I, *BamH* I is not blocked by *dam* methylation.

Cloned at New England Biolabs, Inc.

Figure 1.5. Restriction enzyme specifications. (Courtesy of New England BioLabs, Inc.)

EXPERIMENT

2 Cloning of Restriction Fragments into pUC Vectors

Introduction

In this experiment, you will clone *Hin*dIII restriction fragments of λ bacteriophage DNA into an *E. coli* vector called pUC 13 and then screen for a desired clone by colony hybridization. The DNA inserted into the vector will be the restriction-mapped fragment.

The plasmid vectors used in these cloning procedures must have the following properties:

1. a bacterial replication origin which confers the ability to replicate in a bacterial host
2. genetic markers—usually these are antibiotic resistance markers; some cloning vehicles have two markers, one of which is used as the cloning site, the other of which is used throughout the cloning as a selectable marker
3. a series of sequences with suitable cloning sites in them

In these cloning procedures, we will make use of the pUC series of vectors (Figure 2.1) which contain two genetic markers: one of these is the β-lactamase gene and the other is the β-galactosidase gene. The β-lactamase gene confers resistance to ampicillin (amp^R). The β-galactosidase gene product confers the ability to metabolize a chromogenic substance called X-Gal (5-bromo-4-chloro-3-indolyl-β-D-galactoside) into a blue pigment. Bacterial colonies (transformants) which contain a functional β-*gal* gene on their plasmids will appear blue; those which contain a nonfunctional β-*gal* gene appear white.

The pUC vectors (Vierra and Messing, 1982; Norrander *et al.*, 1983) are ideal for our cloning purposes because they contain a series of restriction sites within a "linker region" in the β-galactosidase gene. DNA fragments inserted into this linker region

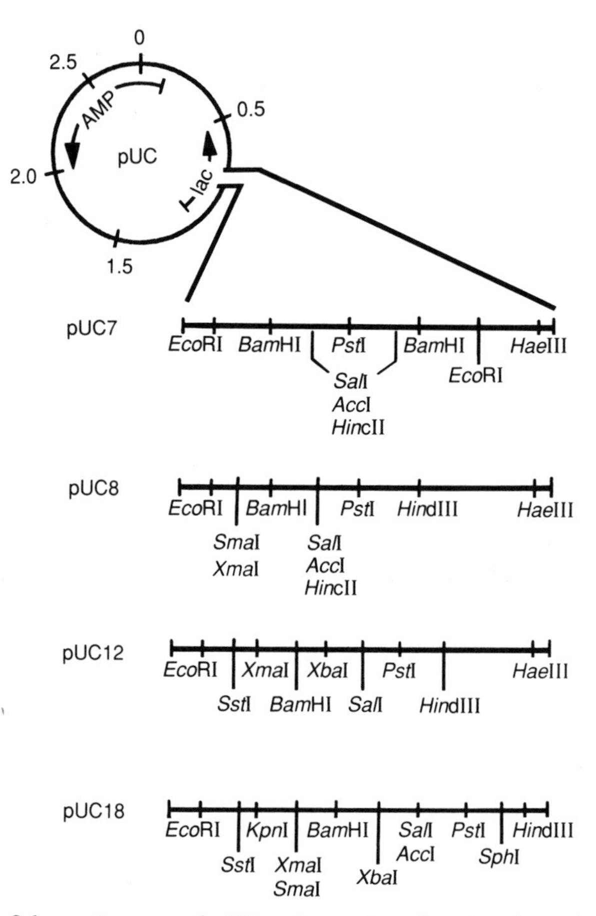

Figure 2.1. Schematic maps of pUC series vectors showing the ordering of unique restriction fragments in the polylinker region.

will inactivate the β-galactosidase gene and will result in the formation of white colonies. [Sometimes small inserts (<300 bp) ligate in-frame with the *β-gal* gene and produce light blue colonies because the β-galactosidase protein is still functional with up to 75 amino acids at its N-terminal end.] The linker regions in the different pUC vectors have slightly different restriction sites (Figure 2.1). The pUC vectors with the paired numbers (pUC 12/13 or pUC 18/19) have the linker regions in reverse orientation relative to the *β-gal* gene. This facilitates the cloning of restriction fragments having different reading frames behind the *β-gal* promoter.

Ligation Reactions

The T4 DNA ligase enzyme catalyzes the covalent joining of certain kinds of DNA fragments to one another. The DNA fragments generated by restriction endonucleases have phosphate groups on their 5′ ends and hydroxyl groups on their 3′ ends (Figure 2.2). If two sticky-ended DNA fragments (with their 5′ phosphate groups intact and "compatible" sticky ends) are mixed, T4 DNA ligase from T4 phage-infected *E. coli* will ligate

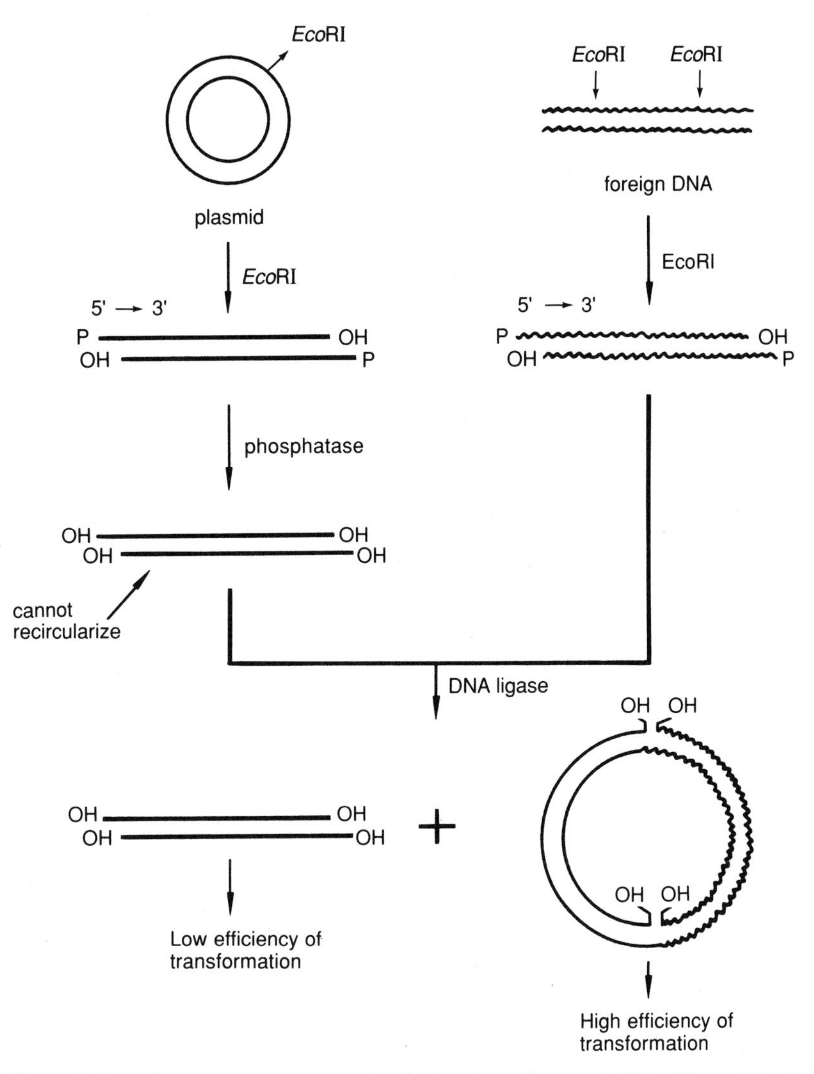

Figure 2.2. Use of phosphatase to prevent recircularization of vector DNA. [From Maniatis *et al.* (1982), p. 15.]

them. Removal of the phosphate groups with bacterial alkaline phosphatase from both kinds of fragments will prevent their joining; removal of phosphate groups from only one of the fragments will allow their joining, but the product will be "nicked" circular DNA which can transform *E. coli* almost as efficiently as closed circular DNA. This type of phosphatase treatment is used in particularly difficult clonings to decrease the number of vector molecules religating on themselves and increase the insertion frequency.

In the example shown in Figure 2.2, the "sticky-ended" DNA molecules being joined to one another must have matching ends (three or four complementary nucleotides). These sticky-ended ligations are much more efficient than the ligation of blunt-ended DNA fragments to one another because the short duplex formed between the complementary nucleotides stabilizes the ends of the DNA for intermolecular ligations. (Check Figure 1.1 to see which enzymes have mutually compatible sticky ends.) In contrast to the specificity of the sticky-ended ligations, blunt-ended ligations can be done much less specifically: any blunt-ended fragment can be ligated to blunt-ended vector DNA. If the same enzyme (i.e., *Sma*I) is not used to generate the blunt ends in the vector and the fragment to be cloned, the blunt-ended restriction site is not reconstructed and the fragment cannot be cut out.

In ligation reactions, it is sometimes essential to eliminate other enzyme activities (*Eco*RI, *Hin*dIII, bacterial alkaline phosphatase) before attempting the ligation reaction. This can be achieved by deproteinizing the DNA fragments with phenol:chloroform (1:1) and precipitating the DNA with 0.2 *M* NaCl and 2 volumes of ethanol before proceeding with the ligation.

Ligation reactions are run at low temperature (4–12°C), below the temperature optimum for the enzyme, because the rate-limiting step in the reaction is the association of sticky or blunt ends with one another. The duplexes formed by sticky ends are very unstable at high temperature and have a longer half-life at low temperature, hence the rate of ligation is faster at lower temperature. Blunt ends have even less affinity for one another and their ligation is aided by very low temperatures (4°C) and the presence of additional ATP in the reaction. For convenience, ligation reactions are carried out at refrigerator temperature (4°C). As with other enzymes, it is important to not shake or vortex the ligase so that it denatures.

Transformation of *E. coli* Cells

Both closed circular and nicked circular DNA molecules can transform "competent" *E. coli* cells with good efficiency. Once inside the bacterial cell, the nicked DNA molecules are ligated by endogenous *E. coli* DNA ligases.

In the transformation process, the ligated DNA samples are added to "competent" *E. coli* cells that contain no plasmids. Competent cells are bacterial cells which have been harvested in log phase and pretreated with $CaCl_2$ so that they efficiently take up DNA. After the ligated plasmid DNA is taken up by the *E. coli,* the *E. coli* are allowed to grow for a period of time in nonselective media so that the antibiotic resistance genes on the plasmid are expressed. Then the bacteria are plated out on selective antibiotic-containing media. The plates are incubated overnight at 37°C so that the individual transformants can grow into distinct colonies. At this point, the plates containing bacteria can be stored in the refrigerator until the colonies are analyzed.

If a mixture of restriction fragments (B′, B″, B‴, etc.) are ligated into the vector fragment (A) and *E. coli* cells are transformed with the mixture of ligated plasmids, then each of the resulting colonies should contain plasmids with the vector DNA and a different inserted fragment of DNA. The efficiency of the *E. coli* transformation decreases as the size of the plasmid DNA increases. Therefore, the proportion of colonies containing a particular DNA fragment will be proportional to the number of fragments to be cloned *and* to the size of the fragment. The higher molecular weight fragments will be cloned much less frequently than the lower molecular weight fragments.

A side reaction in the ligation discussed above is that A, the plasmid vector, can recircularize on itself without inserting any B-type fragment at all. The use of the pUC vectors, with their functional β-galactosidase gene, eliminates many of the problems associated with the vector DNA reclosing on itself, because cutting and religation of the vector DNA without the insertion of any exogenous DNA restores the reading frame of the β-galactosidase gene. When this functional *β-gal* gene is carried in an *E. coli* cell whose endogenous *β-gal* gene has been inactivated (TB-1 cells), β-galactosidase will be produced and the colonies will turn blue on X-Gal plates. DNA fragments inserted into the *β-gal* gene will inactivate the gene and produce white *E. coli* transformants on the X-Gal plates. [Again, small inserts in-frame with the β-galactosidase gene may produce light blue colonies.]

Once you have a group of transformants containing "inserts," the method that you use to identify the correct clone depends greatly on the nature of the cloning experiment. There are two procedures that are generally used: "colony hybridization" or restriction analysis on "minipreps" of the DNA. We will first identify the colony having a particular fragment of λ DNA in it by colony hybridization (Grunstein and Hogness, 1975). Then we will analyze the cloned plasmid by restriction enzyme analysis of a miniprep of the plasmid.

Colony hybridizations are a form of DNA hybridization carried out on nitrocellulose filters. In this procedure, you do not purify the DNA from the bacterial cells, but instead you grow and lyse the *E. coli* colonies right on the nitrocellulose filters. The lysing process is done in such a way that the DNA denatures (the strands separate) and binds directly to the filter. Because the bacteria are grown directly on the filter, it is sometimes referred to as a "colony filter."

The colony containing a particular DNA fragment can be identified by hybridizing a denatured "nick-translated" ^{32}P-labeled DNA probe to the colony filter. The colonies which contain sequences complementary to the probe will bind the radioactive DNA, and when the resulting filters are exposed to X-ray film, the desired colonies will "light up" on the autoradiograph.

Because the ^{32}P-labeled DNA used as probe must not have any homology with the vector DNA (otherwise all the colonies will hybridize to the probe), we will provide you with an unlabeled probe fragment purified from an agarose gel which you can nick translate.

In the nick-translation procedure, DNA is lightly nicked on one strand with DNase I and repaired with DNA polymerase I in the presence of a ^{32}P-labeled dATP ([^{32}P]dATP) and unlabeled dGTP, dCTP, dTTP. After the nick-translation reactions are finished, the reaction mixture is run over a G-100 Sephadex column to remove the [^{32}P]dATP not incorporated into DNA so that it does not randomly stick to the nitrocellulose filters in the subsequent steps. **Wear gloves at all times when handling ^{32}P and check your hands with the Geiger counter immediately after handling the radioactive tubes and before leaving the lab!**

Another Note of Caution: Because this is a recombinant DNA experiment, no mouth pipetting is permitted once the DNA is added to the bacteria. Bacterial spills must be cleaned up with Clorox and your hands should be washed with soap and water at the end of the day. Bacterial plates should be autoclaved at the conclusion of the experiment.

Hybridization Conditions

In general, there are several important parameters that should be considered in performing nucleic acid hybridizations: temperature, solvent, salt concentration, probe concentration, and degree of homology.

The highest rate of nucleic acid hybridization occurs at about 25°C below the temperature, T_m, at which the DNA is 50% melted (that is, converted from double-stranded DNA to denatured single-stranded DNA). For this reason, most hybridizations are carried out 25°C below the T_m of the DNA duplex (T_m − 25°C). In order to select for stable hybrids, hybridization is followed generally by washes at T_m − 15°C. In calculating the T_m for any nucleic acid, one must consider the GC richness of the sequence, the length of the DNA hybrid, the fidelity of base pairing between the probe and the DNA on the filter, and the combination of salt and solvents employed. One chooses hybridization conditions that will preferentially select for certain hybrids in a complex mixture of nucleic acids. All hybridization conditions are based on the formula:

$$T_m = 16.6(\log[Na^+]) + 0.41[(G + C)\%] + 81.5$$

Using this formula, there are several useful relationships to consider when choosing hybridization conditions:

1. In 1× SSC (~0.15 *M* NaCl), at neutral pH and with nucleic acids in the range of 50 to 1000 nucleotides long, the T_m = 69.3°C + 0.41[(G + C)%] (Marmur and Doty, 1962). In 5× SSC (0.83 *M* NaCl) hybridization solutions, the T_m = 80.2°C + 0.41[(G + C)%].
2. T_m is reduced by 0.7°C for each 1% of formamide substituted for water as the solvent (McConaughy *et al.*, 1969).
3. T_m is reduced by 1°C for each 1% increase in the number of mismatched base pairs in a nucleic acid duplex (Bonner *et al.*, 1973).
4. T_m is reduced if the hybridizing regions are very short (<100 nucleotides). The correction for short regions of homology is $500/n$, where n = number of nucleotides in hybridizing regions.

If the G + C content of a DNA probe is not known, a useful approximation for calculating a T_m for single-copy DNA in preliminary experiments is 50% G + C. Thus in 1× SSC, T_m = 69.3°C + (0.41)(50) = 89.8°C. Hybridizations are often set up in solutions of 40–50% formamide because they present less of an evaporation problem and are less harsh on nitrocellulose filters than are aqueous hybridizations. In this hypothetical example

$$T_m\ f = T_m\ H_2O - (0.7°C)(\%\ formamide) = 89.8°C - (0.7°C)(50) = 54.8°C$$

Since nucleic acid hybrids are more stable in higher concentrations of salt, most hybridizations are carried out in 4× or 5× SSC. In 5× SSC, the parameters of this hypothetical hybridization change slightly.

$$T_m = 80.2°C + 0.41[(G + C)\%] = 80.2°C + 0.41(50) = 100.7°C$$
$$T_m\ f = T_m\ H_2O - (0.7°C)(\%\ formamide) = 100.7°C - (0.7)(50) = 65.7°C$$

For this hybridization, the temperature at which maximum hybridization rate occurs ($T_m - 25°C$) would be 40.7°C.

A number of additives are often made to hybridization solutions in the hope of decreasing any nonspecific binding of ^{32}P-labeled probe DNA to nitrocellulose filters. These are generally substances that act as surfactants [bovine serum albumin (BSA), sodium dodecyl sulfate (SDS)] or polymers (Ficoll, polyvinylpyrrolidone) or heterologous DNA (denatured calf thymus DNA) that saturate "sticky" sites on nitrocellulose filters. More detailed considerations of hybridization conditions are given in Maniatis *et al.* (1982) and Schleif and Wensink (1981).

Following hybridization, the nonhybridized probe DNA is removed by washing the filters in low salt solutions, usually at increased temperature, in order to select the most stable hybrids formed. Normally a criterion of $T_m - 5°C$ to $T_m - 15°C$ is chosen. Obviously, the more stringently the filters are washed, the lower the nonspecific background and the more positive the hybridization signal will be. If you are uncertain of the degree of homology between the probe and your DNA, you can perform a series of test hybridizations and washes using DNA blotted onto nitrocellulose to determine the optimal washing conditions that produce the highest signal intensity and lowest background.

Protocols

A. Restriction Digestion and Ligation of DNA into pUC Vectors

1. To a microfuge tube add

   2 μl 0.5 μg/μl λ DNA
   1 μl 0.1 μg/μl pUC 13 DNA
   5 μl 10× *Hin*dIII buffer
   41 μl sterile water
   1 μl 1 U/μl *Hin*dIII

   50 μl

2. Incubate 30 minutes at 37°C.
3. Add 5 μl 2 *M* NaCl (final concentration = 0.2 *M* NaCl) and 100 μl ethanol (2 volumes).
4. Freeze on dry ice (−70°C) for 20 minutes.
5. Centrifuge for 5 minutes in Eppendorf centrifuge to pellet DNA; discard supernatant.
6. Cover centrifuge tube with Parafilm; punch hole in Parafilm; dry in desiccator for 5 minutes to remove all of the ethanol.
7. To set up the ligation, add the following ingredients to the tube*:

*Student Note □ Make sure you pipet each of these to the bottom of the tube; do not vortex or centrifuge after you have added the ligase to the ligation mix.

3 μl 10× T4 DNA ligase buffer
1 μl 100 m*M* ATP†
25 μl sterile water
1 μl 1 U/μl T4 DNA ligase

30 μl

†Student Note □ ATP breaks down at room temperature and with repeated freezing and thawing, so put it on ice once it is thawed.

8. Incubate ligation mixture overnight in refrigerator (4°C) or for 1 hour at 15°C.

Procedure for Teaching Assistants

Teaching assistants should make these the day before.

Preparation of Competent Cells

1. Inoculate 100 ml YT broth with 1 ml of a fresh TB-1 overnight; shake at 37°C until OD_{650} = 0.2; chill on ice for 5 minutes and keep cells cold for the rest of the preparation.
2. Harvest TB-1 cells by centrifuging at 5000 rpm in HB-4 rotor (3000 *g*) for 5 minutes; pour off supernatants.
3. Resuspend pellets in 5 ml *ice-cold* 0.1 *M* $CaCl_2$ by agitating while in ice bath; recentrifuge at 5000 rpm for 5 minutes in HB-4 rotor.
4. Resuspend cells in 1 ml cold 0.1 *M* $CaCl_2$ and store on ice for at least 2 hours (preferably overnight). (Each group will need 600 μl competent cells—200 μl per each of two transformations plus some to plate out without added DNA.)

B. Transformation

1. Set up two transformations
   a. real transformation: add 1 μl ligation reaction to 200 μl competent cells
   b. control transformation*: add 1 μl 1:10 dilution of pUC 13 stock (0.1 μg/μl) to 200 μl competent cells (0.01 μg pUC DNA)

*Student Note ▫ The other control that should be done when using a new batch of competent cells is to plate out competent cells alone with no plasmid DNA. No colonies should result from this plating because the TB-1 cells contain no plasmid DNA.

2. Incubate both transformations at 0°C (on ice) for 20 minutes.
3. Incubate 90 seconds at 42°C.
4. Add 1 ml sterile YT broth; incubate at 37°C for an additional 45 minutes.
5. Plating
   a. plate 200 μl wet cells transformed with ligation reaction onto each of three XG-Amp plates (40 μg/liter X-Gal, 0.2 g/liter ampicillin)
   b. plate 200 μl wet cells transformed with pUC 13 vector DNA onto one XG-Amp plate; discard the rest of the transformation reaction
6. Incubate all plates overnight upside down at 37°C; refrigerate plates to prevent colonies from getting too big. (Colonies with intact pUC 13 will be blue. Colonies with inserts in β-galactosidase gene will be white.)

C. Colony Hybridization

1. Set up a master plate of your white colonies by picking 48 colonies onto a new XG-Amp plate. Use sterile toothpicks to transfer colonies onto a 6 colony × 8 colony grid set up on the master plate.
2. As you inoculate these colonies onto the "master plate" you should also replica plate them onto an Amp plate which has a circular nitrocellulose filter placed on it with forceps.

3. Turn both plates upside down and incubate them overnight at 37°C.
4. Line three plastic or Pyrex dishes with three layers of Whatman 3MM paper. Saturate the paper with the following solutions in separate dishes:

   a. 0.5 *M* NaOH
   b. 1 *M* Tris (pH 7.4)
   c. 0.5 *M* Tris (pH 7.4), 1 *M* NaCl

   Allow the colonies to absorb first to the paper soaked with NaOH solution (colony side up) for 5 minutes at room temperature.* Then transfer the filters to the Tris (pH 7.4) filter pad for 5 minutes and finally to the Tris (pH 7.4), NaCl filter pad for 5 minutes. (The NaOH solution denatures and binds the DNA to the filter; the Tris solutions neutralize the filters.)

*Student Notes

▫ Remove bubbles underneath the nitrocellulose filter by picking up the filter and laying it down again.
▫ Filter papers should be wet but not submersed in NaOH or Tris solutions. The solutions should not float over the top of the nitrocellulose filters!

5. Dry filters at room temperature on paper towels for 30 minutes (colony side up). Place filters in a 75°C oven for 1 hour in order to bake the DNA onto the filters.
6. Prehybridize the filters (to prevent the ^{32}P-labeled probe from nonspecifically binding to the filters) by placing two to four filters in a seal-a-meal bag; add 5 ml solution (5× SSC, 0.2% sarkosyl, 1× Denhardt's solution, 50% formamide) per filter using a 10-ml syringe to add the solution through a corner of the bag; seal bags; check for leaks; submerse bags in a 42°C water bath for at least 8 hours.

   Now we will skip to the nick-translation procedure since this can be done while your filters are baking. The teaching assistant will provide you with 1 μg of purified 1.96-kb and 2.26-kb *Hin*dIII λ DNA fragments (mixed together) so that you can nick translate them and then use the probe to detect the 1.96-kb and 2.26-kb *Hin*dIII fragments cloned in your pUC vector. The nick-translation procedure should be done behind an isotope shield. **Wear gloves! Read notes on radioisotope usuage in Experiment 3.**
7. Set up the nick-translation reaction as follows:

10 μl DNA (1–5 μg CsCl-purified DNA)
10 μl 10× nick-translation buffer [500 m*M* Tris (pH 7.8), 50 m*M* $MgCl_2$, 1 mg/ml BSA, 10 μ*M* 2-mercaptoethanol]
2 μl 20 m*M* cold dNTP stock (dCTP, dGTP, dTTP)
2 μl [^{32}P]dATP (10 μCi/μl, 3000 Ci/mmol)
74 μl H_2O
1 μl 1/10,000 dilution of 1 mg/ml stock DNase (dilute with water immediately before use and keep on ice because DNase I denatures when it is so dilute)
1 μl DNA polymerase I-holoenzyme*

100-μl reaction

*Student Note ▫ As with all enzyme reactions the enzyme should be added last and the reaction should not be vortexed after adding enzyme.

8. Incubate for 1 hour at 15°C in an ice bucket water bath behind the isotope shield.

9. Add 20 μl nick-translation "stop" buffer [0.2 *M* EDTA (pH 7.5), 1% SDS, 40 mg/ml blue dextran, 0.1 mg/ml bromphenol blue]†

†Student Note ▫ The EDTA and the SDS inactivate DNA polymerase I and DNase, the blue dextran acts as a tracking dye for the high molecular weight DNA, and the bromphenol blue acts as a tracking dye for the small unincorporated nucleotides. If you are short on time, the nick translation can be frozen after adding the stop buffer until there is time to run the column.

10. Set up a 4-ml G-100 Sephadex column to separate the ^{32}P-labeled oligonucleotides from the unincorporated [^{32}P]dATP. Stopper a plastic 5-ml pipet (with its top cut off) with siliconized glass wool; fill pipet with TE buffer [TE buffer = 10 m*M* Tris (pH 7.4), 1 m*M* EDTA] and add G-100 Sephadex preswollen in TE buffer until column bed is 4 ml. Add buffer to the top of the Sephadex column so that it doesn't run dry.

11. Layer the 120 μl of the nick-translation reaction onto the top of the Sephadex; after it has adsorbed to the column, start eluting the DNA with sterile TE buffer; collect the first blue peak that comes off the column; this excluded peak is the dextran blue peak and contains the incorporated radioactive nucleotides. The second blue band, the bromphenol blue peak, contains the small, unincorporated nucleotides and should be retained on the column. **Dispose of the column in the radioactive waste area!**

12. Count 1 μl of the first blue peak on a paper circle in a scintillation vial using the ^{3}H channel. (This type of ^{32}P detection without scintillation fluid is referred to as Cerenkoff counting. There should be 100,000 cpm/μl when 2–5 μg DNA are used for nick translation.)
13. After the filters have prehybridized, cut a corner off the seal-a-meal bag and squeeze excess liquid out of the bag.
14. Hybridize your filters with the nick-translated DNA by putting enough ^{32}P counts for 200,000 cpm/filter into 1 ml sterile water; boil the probe for 2–3 minutes in a boiling water bath to separate the DNA strands.
15. Mix the heated probe with the hybridization solution (outlined below) and add to the seal-a-meal bag containing filters; get rid of as many bubbles as possible; reseal the bag.

    Set up each bag of filters with 15 ml hybridization volume as follows:

    7.5 ml 10× SSC, 0.4% sarkosyl, 2× Denhardt's solution
    6.0 ml formamide
    0.5 ml H_2O
    1.0 ml ^{32}P nick-translated DNA in H_2O (approximately 200,000 cpm/filter)
    ______
    15.0 ml

16. Hybridize filters for 12–20 hours in a water bath set at a temperature adequate for the desired level of stringency (usually 35–42°C if there is 40% formamide in the hybridization buffer).
17. Wash off the unhybridized probe by opening up the bag and removing the filters. **Wear gloves!** Float the filters in a Pyrex baking dish with 150 ml 2× SSC, 0.5% sarkosyl; put Pyrex dish in 50°C water bath and wash for 30 minutes; pour off wash solution; add 150 ml 2× SSC buffer and wash for 15 minutes at 50°C; pour off buffer and repeat wash twice for 15 minutes each time (total of four washes). If possible, gently agitate on the shaker table. Dry filters on a paper towel at room temperature.
18. Scotch tape filters onto a small sheet of filter paper. Mark the edges of the filters asymmetrically with ^{32}P-labeled India ink (ink plus a little [^{32}P]dATP).
19. Put Saran wrap over filters so they don't stick to the film. Put them on XAR-5 film and expose in −70°C freezer for 1 day.

20. Take filters off film and develop X-ray. At this point you should have at least a few colonies that have hybridized with your probe. If not, find someone in class who has good positive colonies that you can "miniprep."
21. Inoculate small YT cultures as outlined at beginning of the "miniprep" method and grow cultures overnight at 37°C.

D. Miniprep Purification of DNA

You should pick two colonies that hybridize with your DNA probe and evaluate them by restriction analysis. There are two options in the miniprep procedure given below. The one on the left is used for loading uncut DNA directly onto gels; the one on the right is used for isolating DNA to restriction cut. (You will do the one on the right.) Sometimes the procedure shown in the middle, without any phenol:chloroform extraction, is used to isopropanol-precipitate plasmid DNA rapidly for restriction cutting. DNA prepared this way is not as pure as phenol-extracted DNA so some restriction enzymes won't cut it.

1. Using sterile toothpicks, inoculate 1 ml of YT broth containing 200 μg/ml ampicillin (in a 1.5-ml Eppendorf tube) with each of the two colonies. Put on an extra cap that has a hole punched in it.
2. Incubate the minicultures overnight at 37°C or until the next class starts.
3. Take caps with holes off culture tubes; replace with regular caps (without holes); pellet bacteria by centrifuging in Eppendorf centrifuge for 2 minutes.
4. Resuspend cells in 200 μl solution [8% sucrose, 5% Triton, 50 m*M* EDTA, 50 m*M* Tris (pH 8.0)].
5. Add 10 μl 10 mg/ml lysozyme (0.01 g/ml) and *immediately* put in boiling water bath for 1 minute.
6. Centrifuge for 10 minutes in Eppendorf centrifuge to pellet chromosomal DNA; pour supernatant into new Eppendorf tube.

Miniprep for Sizing Uncut Plasmids	Miniprep for Rapid Restriction Cuts	Miniprep with Phenol Extraction
7. Mix 15 μl supernatant with 10 μl 10 *M* urea loading dye	7. Add 400 μl isopropanol; chill on ice or at −20°C for 30 minutes	7. Add 200 μl phenol:chloroform (1:1)

continued on next page

8. Load entire sample into one well on a 0.8% agarose gel	**8.** Centrifuge for 10 minutes in Eppendorf centrifuge; discard supernatant	**8.** Vortex; centrifuge 2 minutes in Eppendorf centrifuge
	9. Dry pellet and resuspend in 50 μl sterile water	**9.** Transfer aqueous phase to new Eppendorf tube; add 20 μl 2 *M* NaCl and 500 μl ethanol; chill at −70°C (on dry ice) for 20 minutes
	Load 3 μl uncut plasmid/well on 0.6% agarose gel	**10.** Centrifuge for 10 minutes in Eppendorf centrifuge; discard supernatant
	Restriction cut 5 μl plasmid in a 50-μl reaction; add 10 μl loading dye and load 20 μl/well on 0.8–1.2% agarose gel	**11.** Add 50 μl 0.2 *M* NaCl and 100 μl ethanol to DNA; chill on dry ice for 10 minutes; centrifuge for 5 minutes in Eppendorf centrifuge; repeat precipitation one more time
		12. Dry pellet and resuspend in 50 μl sterile water
		Load 3 μl uncut plasmid/well on 0.6% agarose gel

continued on next page

or restriction cut 10 μl plasmid in a 50-μl reaction; add 10 μl loading dye and load 20 μl/well on 0.8–1.2% agarose gel

E. Restriction Analysis of DNA from Minipreps

The purpose of these restriction digestions is to determine if you have cloned the 1.96-kb or the 2.26-kb *Hin*dIII restriction fragment from λ DNA. If you have cloned one of these fragments, then when you cut your minipreps with *Hin*dIII you should release the inserted fragment from the vector and the inserted fragment should line up with the appropriate *Hin*dIII fragment of your *Hin*dIII-cut λ DNA standard. The inserted fragment can be better mapped by looking at the λ DNA map (Figure 1.2, Experiment 1) and by determining which enzymes might differentiate between the 1.96-kb and the 2.26-kb *Hin*dIII fragments.

Remember that the fragment you cloned into the pUC vector has *Hin*dIII sites on either end so that if you want to map a site relative to one end of the insert you need to cut the DNA with a restriction enzyme that has a unique site at either end of the linker region.

You will use 10 μl of your miniprepped DNA per reaction (so you can do only five different reactions on each sample). In planning your five restrictions use both single and double restrictions as you did in Experiment 1.

1. Set up 50-μl restriction digestions as outlined below:

   5 μl 10× restriction enzyme buffer
   10 μl miniprep DNA
   34 μl sterile water
   1 μl 1 U/μl restriction enzyme
   ______
   50 μl

2. Incubate 30 minutes at 37°C.
3. Add 2 *M* NaCl and/or 1 *M* Tris (pH 7.4) (if needed for additional digestions) and the second enzyme.

4. Incubate an additional 30 minutes at 37°C.
5. Add 10 μl 10 *M* urea loading dye.
6. Load 20 μl of each sample onto a 1% agarose gel. Don't forget to load your *Hin*dIII-cut λ DNA and *Hin*f-cut pBR322 standards!!!
7. Electrophorese gel and stain as before.
8. Construct a restriction map of your clone.

Materials Provided
A. DNA Ligation and Transformation

TB-1 cells	ara, Δ(lac$^-$ proAB), rpsL, ϕ80, lacZ Δ M15, hsdR (rk$^-$, mk$^+$)
XG-Amp plates	5 g yeast extract 8 g bacto-tryptone 5 g NaCl 12 g bacto-agar 1 liter water ——— autoclave allow solution to cool to 60°C add 0.2 g ampicillin and 0.04 g X-Gal mix and pour plates store plates at 4°C
YT broth	5 g yeast extract 8 g bacto-tryptone 5 g NaCl 1 liter water ——— mix and autoclave in 50-ml portions
pUC 13 DNA	0.1 μg/μl
λ DNA	0.5 μg/μl
2 *M* NaCl	11.7 g NaCl in 100 ml sterile water

10× T4 DNA ligase buffer	
500 m*M* Tris (pH 7.8)	5 ml 1 *M* Tris (pH 7.8)
100 m*M* $MgCl_2$	0.20 g $MgCl_2 \cdot 6H_2O$
200 m*M* DTE	0.31 g DTE
10 m*M* ATP	0.06 g ATP
500 μg/ml BSA	5 mg BSA (enzyme grade)
	up to 10 ml with sterile water
100 m*M* ATP	51 mg ATP in 1 ml sterile water
0.1 *M* $CaCl_2$	1.47 g $CaCl_2 \cdot 2H_2O$ in 100 ml sterile water

B. Colony Hybridization

0.5 *M* NaOH	20 g NaOH/liter H_2O
1 *M* Tris (pH 7.4)	121 g Tris base add up to 950 ml H_2O adjust pH to 7.4 with conc. HCl (requires a lot of HCl)
	adjust final volume to 1 liter with distilled water
0.5 *M* Tris (pH 7.4) 1 *M* NaCl	60.5 g Tris base 58.4 g NaCl add up to 950 ml H_2O adjust pH to 7.4 with conc. HCl
	adjust final volume to 1 liter with distilled water
10× SSC	175 g NaCl (1.5 *M*) 88 g sodium citrate (0.15 *M*)
0.4% sarkosyl	8 g sarkosyl
2× Denhardt's solution	
0.04% BSA	0.8 g BSA
0.04% polyvinyl-pyrrolidone	0.8 g polyvinyl-pyrrolidone
0.04% Ficoll	0.8 g Ficoll
	adjust final volume to 2 liters with distilled water

2× SSC 0.5% sarkosyl	100 ml 20× SSC 5 g sarkosyl adjust final volume to 1 liter with distilled water
20× SSC	175 g NaCl 88 g sodium citrate adjust final volume to 1 liter with distilled water

C. Nick Translation

Purified λ DNA fragments	2.26-kb + 1.96-kb *Hin*dIII fragments gel purified by teaching assistants
10× nick-translation buffer	
500 m*M* Tris (pH 7.8)	50 ml 1 *M* Tris (pH 7.8)
50 m*M* $MgCl_2$	1 g $MgCl_2 \cdot 6H_2O$
1 mg/ml BSA	0.1 g BSA (enzyme grade)
10 μM 2-mercaptoethanol	0.1 μl 2-mercaptoethanol
	up to 100 ml with sterile water
1 mg/ml DNase I stock	
20 m*M* dCTP, dGTP, dTTP stock solutions	
Nick-translation stop buffer	
0.2 *M* EDTA (pH 7.5)	0.67 g EDTA
1% SDS	0.1 g SDS
40 mg/ml blue dextran	0.4 g blue dextran
0.1 mg/ml bromphenol blue	1 mg bromphenol blue
	up to 10 ml with sterile water
G-100 Sephadex swollen in TE buffer	5 ml/student
TE buffer	
10 m*M* Tris (pH 7.4)	5 ml 1 *M* Tris (pH 7.4)
1 m*M* EDTA	2.5 ml 0.2 *M* EDTA (pH 7.4)
	up to 500 ml with distilled water autoclave

D. DNA Minipreps

YT broth with 0.2 g ampicillin/liter

Lysis buffer

8% sucrose	16 g sucrose
5% Triton X-100	10 ml Triton X-100
50 m*M* EDTA	50 ml 200 m*M* EDTA (pH 7.5)
50 m*M* Tris (pH 8.0)	10 ml 1 *M* Tris (pH 8.0)
	up to 200 ml with sterile water

10 mg/ml lysozyme (freshly prepared)
Phenol:chloroform (1:1) (saturated with TE buffer)

References

Bonner, T. I., Brenner, D. J., Neufeld, B. R., and Britten, R. J. (1973). Reduction in the rate of DNA reassociation by sequence divergence. *J. Mol. Biol.* **81,** 123–125.

Grunstein, M., and Hogness, D. S. (1975). Colony hybridization: A method for the isolation of cloned DNAs that contain a specific gene. *Proc. Natl. Acad. Sci. U.S.A.* **72,** 3961–3965.

Maniatis, T., Fritsch, E. F., and Sambrook, J. (1982). "Molecular Cloning: A Laboratory Manual," pp. 3–15, 146, 270. Cold Spring Harbor Laboratory, Cold Spring Harbor, New York.

Marmur, J., and Doty, P. (1962). Determination of the base composition of deoxyribonucleic acid from its thermal denaturation temperature. *J. Mol. Biol.* **5,** 109–118.

McConaughy, B. L., Laird, C. B., and McCarthy, B. J. (1969). Nucleic acid association in formamide. *Biochemistry* **8,** 3289–3295.

Norrander, J., Kempe, T., and Messing, J. (1983). Construction of improved M13 vectors using oligo-directed mutagenesis. *Gene* **26,** 101–106.

Rodriguez, R. L., and Tait, R. C. (1983). "Recombinant DNA Techniques: An Introduction," Chapters 6, 7, 8. Addison-Wesley, Reading, Massachusetts.

Schleif, R. F., and Wensink, P. C. (1981). "Practical Methods in Molecular Biology," p. 145. Springer-Verlag, Berlin and New York.

Vierra, J., and Messing, J. (1982). The pUC plasmids, an M13mp7-derived system for insertion mutagenesis and sequencing with synthetic universal primers. *Gene* **19,** 259–268.

EXPERIMENT

3A Preparation of Intact Chloroplasts from Pea

Introduction

Experiments designed to examine the activities and biochemical constituents of a cellular organelle require a method that physically separates the organelle from other subcellular fractions without disrupting its structural integrity. In chloroplasts from higher plants, fixation of CO_2, CO_2-dependent evolution of O_2, and synthesis of chloroplast proteins and nucleic acids are directly related to the presence of an intact chloroplast envelope membrane. This indicates that to be physiologically active chloroplasts must also be structurally intact.

This group of experiments, which focus on the isolation of intact chloroplasts for *in organello* labeling studies, is intended as an introduction to organellar isolation and fractionation, protein labeling, and protein gel electrophoresis.

The cellulose walls of the plant tissues used in these procedures are so rigid that the forces needed to disrupt the cell walls tend to rupture fragile organelles as well unless special care is taken during homogenization. Of particular concern is the large central vacuole, which contains tannins, acids, and hydrolases that can seriously damage the subcellular components. Thus, the first requirement for successful plant cell fractionation is that the cells be disrupted under mild conditions. A number of devices are currently available for cell breakage (reviewed in Price, 1983), but the most versatile tool for disrupting mature plant leaf cells is a mechanical blender (e.g., a Waring blender).

The second requirement for purification of enzymatically active chloroplasts is that the chloroplasts be protected chemically from the vacuolar contents released during the leaf cell disruption. The grinding medium must be sufficiently buffered to stabilize the pH when vacuolar acids are released. In addition, because chloroplasts are osmotically responsive, the medium must contain an osmoticum similar in concentration to the cytoplasm of the cell. Sucrose, first used for this purpose in the 1930s, is still the most commonly employed osmotic support. In spite of its wide use, however, sorbitol and mannitol are less damaging to chloroplast membranes than sucrose. Solutions used for cell fractionation must be kept cold (typically 0 to 4°C) to arrest the activity of hydrolytic enzymes. Bovine serum albumin is often added as a general protectant for membranes (probably acting as a substrate for endogenous proteases and/or binding free fatty acids). Other components of the grinding medium vary depending on the tissue being used. For example, plants such as tomato and soybean, which are high in phenolic compounds, usually require the presence of compounds that bind the phenolics and others that prevent their oxidation. These phenolics, when oxidized, are very detrimental to chloroplast preparations. Typically, polyvinylpyrrolidone or one of its insoluble derivatives is added to bind phenolic derivatives. Reductants, such as ascorbate, 2-mercaptoethanol, or dithiothreitol, can also be added to prevent phenolic oxidation.

Following disruption of the leaf tissue, it is important to free the extracted chloroplasts of other organelles (mitochondria, tonoplasts, etc.) and small tissue fragments. This is accomplished by first filtering the homogenate through several layers of cheese cloth or nylon cloth in order to remove unbroken cells and large pieces of tissue. The green filtrate contains various organelles (nuclei, mitochondria, Golgi, etc.) and some smaller tissue fragments. These can be separated from one another by centrifugation.

All of the most widely employed methods of organelle isolation utilize some form of differential centrifugation. These procedures take advantage of the fact that subcellular organelles sediment at a characteristic rate when subjected to a centrifugal field. The sedimentation rate primarily depends on the size of the particular organelle. The sedimentation coefficient, or *S* value, of an organelle increases with the size of the organelle. Organelles having larger *S* values pellet more quickly at lower speeds than those with small *S* values. Thus, the smaller organelles are removed in the supernatant and recovered by centrifugation at successively higher speeds or for longer times. The advantage of differential centrifugation is that it has a high capacity for rapidly separating organelles. Although differential centrifugation is a valuable tool for obtaining fractions enriched in chloroplasts, two problems limit its usefulness in further purification. The first difficulty is that it cannot resolve chloroplasts and mitochondria, whose sedimentation coefficients fall within similar ranges. The second difficulty with differential centrifugation is that repeated pelleting damages the chloroplast envelope.

Having isolated a crude chloroplast fraction by differential centrifugation, the primary method for separating intact chloroplasts from broken thylakoid membranes is density-gradient centrifugation. This type of centrifugation is the sedimentation of particles (or organelles) in a medium in which the viscosity and density vary as a function of the radius of the centrifuge. When a mixture of subcellular organelles is subjected to a centrifugal field in the presence of a density gradient, it is possible to separate them on the basis of either their density or their size (Brakke, 1951). Separation on the basis of density is referred to as isopycnic (same density) sedimentation; separation on the basis of size is referred to as rate–zonal sedimentation. For more detailed descriptions of centrifugation theory see Price (1983).

A critical variable in density-gradient separation of membrane-bound organelles is the choice of gradient material. Silica sols have been used extensively as density-gradient materials in recent years for plant organelle isolations because they meet most of the criteria for an ideal density-gradient medium: they are freely soluble in water, nonviscous, physiologically and chemically inactive, and have negligible osmotic potential. One of the earliest studies employing silica sol gradients for organelle purification demonstrated that chloroplasts recovered from such gradients were photosynthetically active (Morgenthaler *et al.*, 1974). It was shown subsequently that similarly purified chloroplasts are active in protein synthesis (Morgenthaler and Mendiola-Morgenthaler, 1975) and in the uptake of cytoplasmically synthesized chloroplast precursor proteins (Grossman *et al.*, 1980). Silica sol gradients have also been utilized successfully to isolate plant mitochondria (Yamaya *et al.*, 1984) and peroxisomes (Schwitzguebel and Siegenthaler, 1984).

The experiment outlined here will employ a combination of differential and density-gradient centrifugation to prepare intact chloroplasts from pea seedlings. The proteins in these chloroplast preparations will be examined further by gel electrophoresis in Experiment 3D.

Protocols

A. Isolation of Chloroplasts

(All steps are performed on ice, where possible, using cold reagents)

> **Instructor's Note:** These gradients may be prepared up to 24 hours in advance and stored at 4°C.

1. Pipet 35 ml of Percoll gradient solution into a 50-ml centrifuge tube. Centrifuge the solution at 40,000 *g* (for example, 17,000 rpm in a Beckman JA17 rotor) and 4°C for 40 minutes. In order to conserve centrifuge time, you should plan to spin your gradient solution with those of one or two other lab groups. During the spin, organize the other materials you will need for the chloroplast preparation. Begin harvesting the seedlings (step 2) when the centrifuge timer indicates that there are 10 to 15 minutes remaining of the spin. When the gradients have finished centrifuging, **carefully** remove them from the centrifuge and store at 4°C supported in a test tube rack.

> **Instructor's Note:** The seedlings should be "destarched" by placing them in the dark for about 16 hours prior to starting the experiment. Starch accumulation in chloroplasts will tend to rupture the organelles upon homogenization.

2. Harvest pea seedlings by cutting just above the soil line with a razor blade or scissors. Collect the plant material in a beaker containing ice water. Rinse the leaves briefly, dry on toweling, and then weigh. Collect about 20 g of material.
3. On a glass plate, chop the tissue into small pieces using a razor blade (much as you would chop salad material with a chef's knife). Transfer the pieces to a blender jar containing 200 ml of ice-cold GR medium (for composition see Materials Provided section). Grind the tissue, using two 2-second bursts of the blender with a 2- to 3-second rest between the bursts.
4. Filter the homogenate through four layers of cheese cloth and one layer of Miracloth layered on top of one another in a funnel (arrange the filter so the homogenate passes through the cheese cloth first); collect the filtered homogenate in a 250-ml flask on ice.
5. Distribute the homogenate equally between four 50-ml centrifuge tubes. Discard excess homogenate. Centrifuge the homogenate for 4 minutes at 2500 *g* (3700 rpm in a Beckman JS7.5 rotor) and 4°C.
6. Decant and discard the supernatant liquid. Gently resuspend the crude chloroplast pellets in a total of 10 ml of fresh GR medium. This can be accomplished by repeated aspiration of the pellet in and out of a 10-ml pipet or by adding a

few milliliters of GR medium to each tube and resuspending the chloroplasts with the aid of a fine, camel's hair brush. Save 100 μl of the suspension on ice in a 1.5-ml microfuge tube for chlorophyll and protein estimation.

7. Carefully layer the chloroplast suspension over the Percoll gradient formed in step 1. To do this, aspirate the plastid suspension into a 10-ml pipet; then touch the tip of the pipet to the surface of the gradient and very gently allow the suspension to flow onto the top of the gradient.
8. Centrifuge the gradient at 9000 *g* and 4°C for 15 minutes in a swinging bucket rotor (7000 rpm in Beckman JS7.5, or 9000 rpm in Sorvall HB4). Once again, to conserve centrifuge time, spin your gradient with those of one or two other lab groups.
9. When the centrifuge has come to rest, remove the gradient from the rotor. You should observe two green bands. The upper contains stripped thylakoids (the remains of chloroplasts with broken envelope membranes) and the lower intact chloroplasts. Remove the bands from the gradient using a Pasteur pipet and collect them separately in fresh centrifuge tubes on ice. Be sure to mark the centrifuge tubes to indicate which band is which!
10. Dilute the chloroplast suspensions with about 30 ml of GR medium. Cap the tubes tightly with Parafilm and mix by inverting the tubes several times. Remove the Parafilm and collect the chloroplasts by centrifugation at 2500 *g* and 4°C for 5 minutes.
11. Carefully decant and discard the supernatant liquid. Resuspend the chloroplast pellets in about 10 ml of fresh GR medium. Hold the suspension on ice.
12. Transfer a 100-μl aliquot from each suspension to separate microfuge tubes. Add 800 μl of acetone and 100 μl of distilled H_2O to the microfuge tubes, cap, vortex, and allow to stand 5 minutes on ice. Do the same with the 100-μl aliquot of intact plastids saved in step 6. Centrifuge the extracted chloroplast samples for 5 minutes in the microfuge. Transfer the green supernatants to fresh tubes and save the extracted pellets on ice. Process the intact plastid pellet for estimation of protein according to the Lowry assay described in Experiment 3B.
13. Estimate the chlorophyll in your 80% acetone extracts by determining the A_{663} and A_{645}. Read all absorbances for this

estimation against an 80% acetone blank. If the extract is too concentrated to read the absorbance, dilutions should be made with known volumes of 80% acetone. Calculate the amount of chlorophyll by substituting the values you measured into the following equations:

$$\text{Chl } (\mu\text{g/ml}) = [20.2(A_{645}) + 8.02(A_{663})] \times \text{dilution factor } (1000/100)$$

$$\text{Chl } a/b \text{ ratio} = \frac{(1.27 \times A_{663}) - (0.269 \times A_{645})}{(2.29 \times A_{645}) - (0.468 \times A_{663})}$$

14. After measuring the chlorophyll in the 80% acetone extract and estimating your total recovery of chlorophyll, pellet the intact and broken chloroplast samples by centrifugation at 2500 *g* and 4°C for 5 minutes. Decant and discard the supernatant fluid.
15. Resuspend both pellets of chloroplasts in chloroplast lysis buffer in a final volume that corresponds to a 250 μg/ml chlorophyll concentration. Incubate the suspensions with occasional vortexing for 15 minutes on ice.
16. Divide the lysed intact chloroplasts into 200-μl aliquots (~50 μg chlorophyll) in 1.5-ml microfuge tubes (six to eight aliquots are sufficient). Store one aliquot without further treatment (designate this as "whole chloroplast"). Save one or two 200-μl aliquots of the broken chloroplast suspension (also ~50 μg chlorophyll); discard the remainder.
17. Collect the thylakoid membranes by centrifuging the aliquots for 5 minutes in the microfuge (in the cold room). Transfer the supernatants (stroma) to fresh tubes. Wash the pellet with an equal volume of chloroplast lysis buffer. Centrifuge for 5 minutes and discard the supernatant. Label the tubes appropriately (fraction, name, date, etc.) and store at −70°C.
18. Precipitate one aliquot of stromal extract by adding 800 μl of acetone, mixing well, and incubating on ice for 20 minutes. Collect the precipitated proteins in a microfuge.
19. Analyze the protein content of the acetone-extracted intact chloroplast sample from step 12 and the stromal extract from step 18 by the Lowry procedure.

B. Data Evaluation

Set, in tabular form, the chlorophyll content and percentage recovery, chlorophyll *a*/*b* ratio, protein content, and protein/

chlorophyll mass ratio for each of the "chloroplast" preparations (crude suspension, stripped thylakoids, and purified chloroplasts).

Materials Provided

Pea seedlings, 5 to 7 days old
Plastic beaker, 1000 ml, about half-full of ice water for collecting plant material
8 × 10 inch glass plate
Centrifuge tubes (6–50 ml)
Funnels
Cheese cloth/Miracloth
Waring blender

1 M HEPES/KOH (pH 7.5)	119.1 g HEPES add up to 350 ml H_2O adjust pH to 7.5 with KOH ——— adjust final volume to 500 ml with distilled water
1.75 *M* sorbitol	318.8 g sorbitol ——— adjust final volume to 1 liter with distilled water store at −20°C
0.5 *M* Na_2EDTA	46.53 g $Na_2EDTA \cdot 2H_2O$ add up to 200 ml H_2O adjust the pH to 8.0 with NaOH pellets (about 5) ——— adjust final volume to 250 ml with distilled water
1 *M* $MgCl_2$	10.15 g $MgCl_2 \cdot 6H_2O$ ——— adjust final volume to 50 ml with distilled water
1 *M* $MnCl_2$	9.90 g $MnCl_2 \cdot 4H_2O$ ——— adjust final volume to 50 ml with distilled water

GR medium (~300 ml/group)

0.35 *M* sorbitol	200 ml 1.75 *M* sorbitol
50 m*M* HEPES/KOH	50 ml 1 *M* HEPES/KOH (pH 7.5)
2 m*M* EDTA	4 ml 0.5 *M* Na_2EDTA
1 m*M* $MgCl_2$	1 ml 1 *M* $MgCl_2$
1 m*M* $MnCl_2$	1 ml 1 *M* $MnCl_2$
	adjust final volume to 1 liter with distilled water
	store at 4°C
	just before use, add 1 g sodium ascorbate

PBF–Percoll

	140 ml Percoll
	4.2 g polyethylene glycol 8000
	0.14 g bovine serum albumin
	1.4 g Ficoll 400
	mix ingredients with rapid stirring
	store at 4°C

Percoll gradient solution (35 ml/group)

50% PBF–Percoll	140 ml PBF–Percoll
0.35 *M* sorbitol	56 ml 1.75 *M* sorbitol
50 m*M* HEPES/KOH	14 ml 1 *M* HEPES/KOH (pH 7.5)
1 m*M* $MgCl_2$	0.28 ml 1 *M* $MgCl_2$
1 m*M* $MnCl_2$	0.28 ml 1 *M* $MnCl_2$
2 m*M* Na_2EDTA	1.12 ml 0.5 *M* Na_2EDTA
	68 ml H_2O
	store at 4°C
	just before use add 140 mg sodium ascorbate

80% (v/v) acetone

	800 ml acetone
	200 ml H_2O
	store in a glass-stoppered bottle at room temperature

Chloroplast lysis buffer

62.5 m*M* Tris–HCl (pH 7.5)	6.25 ml 1 *M* Tris–HCl (pH 7.5)
2 m*M* $MgCl_2$	0.2 ml 1 *M* $MgCl_2$
	adjust final volume to 100 ml with distilled water

References

Brakke, M. K. (1951). Density gradient centrifugation: A new separation technique. *J. Am. Chem. Soc.* **73,** 1847–1848.

Grossman, A., Bartlett, S., and Chua, N.-H. (1980). Energy-dependent uptake of cytoplasmically synthesized polypeptides by chloroplasts. *Nature* (*London*) **285,** 625.

Lilley, R. McC., Fitzgerald, M. P., Rienits, K. G., and Walker, D. A. (1975). Criteria of intactness and the photosynthetic activity of spinach chloroplast preparations. *New Phytol.* **75,** 1–10.

Morganthaler, J.-J., and Mendiola-Morganthaler, L. R. (1976). Synthesis of soluble, thylakoid, and envelope membrane proteins by spinach chloroplasts purified from gradients. *Arch. Biochem. Biophys.* **172,** 51–58.

Morgenthaler, J.-J., Price, C. A., Robinson, J. M., and Gibbs, M. (1974). Photosynthetic activity of spinach chloroplasts after isopycnic centrifugation in gradients of silica. *Plant Physiol.* **54,** 532–534.

Price, C. A. (1983). *In* "Isolation of Membranes and Organelles from Plant Cells" (J. L. Hall and A. L. Moore, eds.), pp. 1–24. Academic Press, London.

Schwitzguebel, J.-P., and Siegenthaler, P.-A. (1984). Purification of peroxisomes and mitochondria from spinach leaf by percoll gradient centrifugation. *Plant Physiol.* **75,** 670–674.

Yamaya, T., Oaks, A., and Matsumoto, H. (1984). Stimulation of mitochondrial calcium uptake by light during growth of corn shoots. *Plant Physiol.* **75,** 773–777.

EXPERIMENT

3B Lowry Assay for Protein Determination

Introduction

This assay (Lowry *et al.*, 1951) is sensitive and very reproducible. A calibration curve with a protein such as lysozyme must be determined in parallel with the unknown samples in order to accurately estimate their concentration. The response of the assay is usually nonlinear (i.e., the calibration curve will deviate from a strict Beer's law relationship). The assay responds to tyrosine and tryptophan residues, so different proteins, depending on their aromatic amino acid content, can give very different responses when equal masses are assayed.

In addition, many commonly used buffers interfere with this assay. Thus, samples are normally precipitated with acetone or TCA prior to assaying. In spite of these drawbacks, it is still routinely used because of its reliability.

Protocol

1. To the pellets derived from 80% acetone extraction, add 1 ml of ethanol:ether (1:1). Cap tubes tightly, vortex, and collect protein residue in the microfuge for 5 minutes. Discard the supernatant.
2. Reextract the pellets with 1 ml of ether as described above. Collect proteins by spinning for 10 minutes in the microfuge. Discard the ether and allow the pellets to dry for 5 to 10 minutes in the fume hood.
3. Based on your estimates of chlorophyll in the original samples, dissolve each pellet in a volume of 0.1 *N* NaOH, which adjusts the original concentration of chlorophyll to 1 μg chlorophyll/μl. Be certain to record the exact volume used.
4. Take duplicate aliquots equivalent to 3 and 10 μg chlorophyll from each sample and transfer them to a series of 13 ×

100 mm tubes. Construct a series of lysozyme standards, in duplicate, ranging from 10 to 200 μg of protein. Adjust the volume of all tubes to 500 μl with 0.1 *N* NaOH.

5. Prepare the Lowry assay reagents as follows:

Reagent A
- 2 ml $CuSO_4$ solution
- 2 ml NaK tartrate solution
- 96 ml Na_2CO_3 solution

Reagent B
- 5 ml phenol reagent
- 5 ml H_2O

6. To each protein sample and standard, add 5 ml of reagent A from step 5. Mix well and hold 10 minutes at room temperature.

7. To each sample, add 0.5 ml of reagent B. Vortex or mix vigorously for 10 seconds. Incubate 30 minutes at room temperature.

8. Read the A_{650} of each sample and BSA standard. Construct a standard curve of lysozyme content (in micrograms) vs A_{650}, and interpolate the concentrations of the unknown samples from the graph.

Materials Provided

Material	Preparation
2% (w/v) $CuSO_4$	2 g $CuSO_4$ / adjust volume to 100 ml with distilled H_2O
4% (w/v) NaK tartrate	4 g NaK tartrate / adjust volume to 100 ml with distilled H_2O
3% (w/v) Na_2CO_3 in 0.1 *N* NaOH	30 g Na_2CO_3, 4 g NaOH / adjust volume to 1 liter with distilled H_2O
Phenol reagent (Folin–Ciocalteau)	
0.1 *N* NaOH	4 g NaOH / adjust volume to 1 liter with distilled H_2O

Lysozyme standard (1 mg/ml)	100 mg lysozyme
	adjust volume to 10 ml with distilled H_2O calculate exact concentration from $E^{1\%}_{282\ nm} = 26.4$ adjust volume to give 1 mg/ml lysozyme (final concentration) and adjust NaOH concentration to 0.1 *N* NaOH store in aliquots at −20°C

Reference

Lowry, O. H., Rosebrough, N. J., Farr, A. L., and Randall, R. J. (1951). Protein measurements with the Folin phenol reagent. *J. Biol. Chem.* **193,** 265.

EXPERIMENT

3C Protein Synthesis by Isolated Pea Chloroplasts

Introduction

Chloroplasts and mitochondria contain entire genetic systems that are biochemically distinct from the one found in the cytoplasm; the components of the organellar systems are similar to those of bacterial cells (reviewed in Ellis, 1981). The term "genetic system" as used here designates the four linked components required to express genetic information: DNA, DNA polymerase, RNA polymerase, and the protein-synthesizing apparatus. Isolated, intact chloroplasts are particularly active in protein synthesis *in vitro* and have been used to identify specific products of plastid translation (Fish and Jagendorf, 1984), as well as to investigate posttranslational protein transport (Grossman *et al.*, 1980), processing, assembly (Bloom *et al.*, 1980), and translational regulation (Mullet *et al.*, 1986; Cushman and Price, 1986).

In this set of experiments, you will label chloroplast proteins *in organello* in intact chloroplasts isolated from pea seedlings. The labeled proteins will then be fractionated by electrophoresis in acrylamide gels and detected by fluorography.

Note that all steps in this protocol should be carried out under "sterile" conditions and that radioisotopes will be used. Wear gloves to prevent contamination of glassware and solutions with "finger nucleases" and bacteria, and contamination of yourself with radioactivity. Unless noted otherwise, all steps should be carried out on ice, with ice-cold reagents.

Special Precautions for Use of Radioisotopes

1. When working with radioactive material, wear lab coats (these should always be left in the laboratory) and disposable vinyl gloves to minimize the chance of personal contamination. These items will be available in the laboratory.

2. Do not pipet by mouth. Most of the operations will employ very small volumes of liquid that must be dispensed with automatic pipetters. If standard pipets must be used, fill them by suction with a pipet-filling device.
3. Work on a square of absorbent paper backed with polyethylene to prevent contamination of the lab bench. Keep all work items on this paper. Mark one corner and designate it as a "hot area." A small radioactive trash container should be kept there, as well as all items that are suspected to be contaminated (a small, plastic weighing boat works well for this purpose).
4. Put contaminated, disposable items (e.g., pipet tips, microfuge tubes) into the plastic radioactive trash in your radioactive "hot area." At the end of the period, put this trash and any other contaminated disposable items into the trash can marked "Radioactive Waste." The instructors will indicate the locations of waste containers prior to dispensing any radioactive materials.
5. Put contaminated glassware (e.g., the beaker used for the TCA precipitation assay) in a specially marked dishpan containing Isoclean, a special detergent for decontaminating radioactive labware.
6. After completing work, check your hands and the lab bench for contamination. Use the lab monitor set at its most sensitive setting. It is a good idea to work using the "buddy" system: one lab partner handles the radioactive material; the other lab partner remains "cold" to perform monitoring and handling of nonradioactive solutions. It is also a good idea to wash your hands when you are finished handling radioactive materials, even if no contamination is detected.
7. The organic solvents in liquid scintillation fluid are flammable and toxic when inhaled. Use scintillation fluid in the hood. Cap scintillation vials tightly before leaving the hood area.

Protocol

1. Chloroplasts, isolated from pea, will be prepared by the teaching assistant for this exercise. The chloroplasts are suspended in the resuspension buffer (composition given in the Materials Provided section). The chlorophyll concentration will have been measured by the teaching assistant and ad-

justed to 1 mg/ml. Obtain an aliquot corresponding to 90–100 μg of chlorophyll. This should be removed under **sterile** conditions and held in a microfuge tube on ice while the reaction tubes are set up. While the reactions are being assembled, one member of each team should assemble the reagents and materials needed for a TCA-precipitation assay (these components will be needed as soon as the plastid protein synthesis experiment begins).

2. Number a series of six 13 × 100 mm test tubes. Place the tubes on ice and add the following solutions to each, being careful to pipet all components to the bottom of the tube. Wrap tube number 2 in aluminum foil. It will serve as a dark control.

Component (volume, in μl)	Tube number					
	1	2	3	4	5	6
Resuspension buffer	40	→	→	→	→	400
Compensating buffer	2.5	→	→	→	→	25
[^{35}S]Methionine	1.5	→	→	→	→	15
Chloramphenicol	—	—	1	—	—	—
Cycloheximide	—	—	—	1	—	—
FCCP	—	—	—	—	1	—
Ethanol	1	1	—	—	—	—
H_2O	—	—	—	—	—	10
Chloroplasts (1 mg Chl/ml)	5	→	→	→	→	50

Add all components except the chloroplasts, cover tubes with Parafilm, mix briefly by gently vortexing, and return the tubes to the ice bath. Then add the chloroplast suspension to each tube, mix again, and remove duplicate 10-μl aliquots from tube number 6 as a zero-time control for the TCA-precipitation assay. Transfer the tubes to the illuminated water bath and allow them to incubate at 25°C for 45 minutes. At 5- to 10-minute intervals, the tubes should be gently agitated to resuspend any chloroplasts that may have settled.

3. Remove the tubes from the water bath and place on ice. Take duplicate 10-μl aliquots from each assay tube and spot the aliquots on a numbered series of 1-cm squares of Whatman 3MM filter paper. After spotting the samples, the filters are collected in a beaker containing ice-cold 10% TCA. One member of each team should process the filters as described in the TCA assay protocol.

4. To separate the labeled stromal and thylakoid proteins, another member of the group should process the chloroplasts remaining in tube number 6 as follows:
   a. Transfer the chloroplasts to a 15-ml centrifuge tube containing 5 ml of resuspension medium, swirl or gently vortex the tube to mix, and pellet the chloroplasts by centrifuging them at 2500 *g* for 5 minutes at 4°C
   b. Decant the supernatant liquid into the radioactive waste. Take care not to disturb the chloroplast pellet during this step
   c. Repeat the washing step once. Again, discard the supernatant in the radioactive waste
   d. Add 500 μl of chloroplast lysis buffer to the pellet; resuspend the organelles with the aid of a micropipetter, and incubate the suspension on ice for 15 minutes
   e. Transfer the lysed plastid suspension to two 1.5-ml microfuge tubes (250 μl/tube), and collect the thylakoid membranes by centrifugation in a microfuge for 15 minutes
   f. Carefully remove the supernatants and transfer them to fresh 1.5-ml microfuge tubes (on ice). Label the tubes containing the thylakoid pellets with your name and date, and store them at −80°C (they contain the equivalent of about 25 μg of chlorophyll each)
   g. Add 1 ml of cold acetone to each tube containing the soluble proteins. Cap the tubes tightly and incubate them for 30 minutes at −20°C
   h. Collect the precipitated soluble proteins by centrifuging for 15 minutes in a microfuge
   i. Decant the aqueous acetone into the radioactive waste, but be careful not to disturb the pellets! Wash the pellets by adding about 500 μl diethyl ether to each tube, capping tightly, and vortexing vigorously for 30 to 60 seconds. Collect the proteins in a microfuge (10-minute spin)
   j. Decant and discard the ether wash. Allow the pellets to air dry in the fume hood for about 10 minutes. Cap the tubes tightly, label them, and store at −80°C. Each of these pellets contains soluble proteins derived from chloroplasts that originally contained about 25 μg of chlorophyll
5. From the data obtained from the TCA-precipitation assay, calculate the average cpm incorporated into protein by the chloroplasts in each of the treatments. It will be helpful to set these in tabular form, with the value for each treatment calculated as a percentage of the value obtained for tube number 1

(chloroplasts incubated in the light with no added inhibitors). In your laboratory report, interpret the results of this data. [*Hint:* what process(es) does each inhibitor affect?]

6. During the next laboratory period, the translation products fractionating into the stroma and thylakoid membranes will be examined by SDS–gel electrophoresis and fluorography. For this experiment, set up 12.5% acrylamide gels as outlined in Experiment 3D; load with molecular weight markers and the equivalent of 5 μg chlorophyll/lane for your labeled stromal and thylakoid protein fractions. Reaction center polypeptides can be visualized on your gels if one sample of the thylakoid proteins is kept on ice and electrophoresed without heating. Under these conditions, some of the chlorophyll should remain associated with the PSI (and occasionally PSII) reaction center complex and produce a "green" high molecular weight band on your gels.

Materials Provided

Resuspension buffer	
375 m*M* sorbitol	21 ml 1.75 *M* sorbitol
35 m*M* HEPES/KOH (pH 8.3)	3.5 ml 1 *M* HEPES/KOH (pH 8.3)
0.96 m*M* dithiothreitol (DTT)	15 mg DTT (added just before use)
	adjust final volume to 100 ml
	freeze at −70°C in 1-ml aliquots for class
Compensating buffer	
7.5 m*M* $NaPO_4$ (pH 7.5)	1.88 ml 0.2 *M* $NaPO_4$ (pH 7.5)
5 m*M* $MgCl_2$	125 μl 2 *M* $MgCl_2$
5 m*M* all amino acids except methionine	10 ml 25 m*M* amino acid stock (−Met)
	adjust final volume to 50 ml
	freeze at −70°C in 500-μl aliquots for class

[^{35}S]Methionine	~1 mCi/μmol, 14 μCi/μl, 25 μl/group
Chloramphenicol	5 mg/ml in ethanol
Cycloheximide	5 mg/ml in ethanol
FCCP (*p*-fluoromethoxycarbonylcyanide phenylhydrazone)	500 μg/ml in ethanol
Diethyl ether	
Lysis buffer	
35 m*M* HEPES/KOH (pH 8.3)	7 ml 1 *M* HEPES/KOH (pH 8.3)
4 m*M* $MgCl_2$	0.8 ml 1 *M* $MgCl_2$
	adjust final volume to 200 ml

References

Bloom, M. V., Milos, P., and Roy, H. (1980). Light-dependent assembly of ribulose-1,5-bisphosphate carboxylase. *Proc. Natl. Acad. Sci. U.S.A.* **80,** 1013–1017.

Cushman, J. C., and Price, C. A. (1986). Synthesis and turnover of proteins in proplastids and chloroplasts of *Euglena gracilis. Plant Physiol.* **82,** 972–977.

Ellis, R. J. (1981). Chloroplast proteins: Synthesis, transport and assembly. *Annu. Rev. Plant Physiol.* **32,** 111–137.

Fish, L. E., and Jagendorf, A. T. (1984). High rates of protein synthesis by isolated chloroplasts. *Plant Physiol.* **70,** 1107–1114.

Grossman, A., Bartlett, S., and Chua, N.-H. (1980). Energy-dependent uptake of cytoplasmically synthesized polypeptides by chloroplasts. *Nature (London)* **285,** 625–628.

Mullet, J. E., Klein, R. R., and Grossman, A. R. (1986). Optimization of protein synthesis in isolated higher plant chloroplasts. *Eur. J. Biochem.* **155,** 331–338.

EXPERIMENT

3D Separation of Thylakoid and Stromal Proteins by SDS–Gel Electrophoresis

Introduction

Acrylamide gel electrophoresis is the method of choice for fractionating and characterizing mixtures of proteins. Electrophoretic procedures are rapid, and small amounts of protein can be conveniently detected in gels by staining or autoradiography. In addition, it is so simple to perform electrophoretic separations on large numbers of samples that electrophoresis can be used to assay fractions from chromatography columns, radiolabeling experiments, or protein modification experiments.

Electrophoretic migration is inversely proportional to the molecular weight of a protein if the protein is dissociated and denatured with sodium dodecyl sulfate (SDS) and a reducing agent before electrophoresis. SDS is a negatively charged detergent that binds hydrophobically to polypeptide chains. With the addition of reducing agents (2-mercaptoethanol) and heat, all inter- and intrachain disulfide bonds break, leaving the denatured protein fully reduced and separated into individual polypeptide subunits. Most proteins bind a constant amount of SDS per amino acid residue (Tanford *et al.*, 1974). In so doing, they acquire a fixed charge-to-mass ratio. Proteins, under these conditions, differ only in their size. Electrophoresis of such denatured proteins, then, separates them according to size (for exceptions see Neville, 1971). The distance of electrophoretic migration relative to the buffer front is, then, inversely proportional to the $\log_{10}$ of the molecular weight.

Electrophoretic estimation of the sizes of undenatured, multimeric proteins can also be performed in nondenaturing gels (Hedrick and Smith, 1968), but the procedure is much more complex than size estimation in denaturing SDS–gels and will not be considered here.

A wide variety of modifications have evolved from the basic gel electrophoretic technique. The major classes of these modifications have been reviewed previously (Schleif and Wensink, 1981), and are outlined below. More detailed descriptions can be found in Hames and Rickwood (1981).

1. *SDS electrophoresis in a fixed concentration acrylamide gel* (Weber and Osborn, 1969).
2. *SDS electrophoresis with a stacking gel* (Laemmli, 1970). This is the most commonly used variant. A stacking gel is a porous, low-percentage gel having a different buffer composition that is polymerized on top of the separation gel. It permits samples of large volume to be electrophoretically concentrated into a thin band during the initial stages of electrophoresis and results in sharper protein bands.
3. *Gradient gels.* By varying the concentration of acrylamide or cross-linking reagent along the length of the gel, a wider molecular weight range of proteins can be resolved in a single electrophoretic run.
4. *Nondenaturing gels.* These gels, which contain no detergent, are used for electrophoresis of native proteins. These conditions are frequently used to assay for the presence of different isozymes.
5. *Urea gels.* These partially denaturing gels can be used to separate proteins of similar molecular weight by differences in charge-to-mass ratio. Urea can be used in combination with SDS to optimize electrophoretic resolution of particular proteins (e.g., Piccioni *et al.*, 1982).
6. *Isoelectric focusing gels.* These gels utilize charge carriers to generate a pH gradient on electrophoresis. They fractionate proteins on the basis of their isoelectric point.
7. *Two-dimensional gels.* These gels have the widest range of resolution because they fractionate proteins on the basis of two different parameters: their isoelectric point and their molecular weight (O'Farrell, 1975; Hurkman and Tanaka, 1986).
8. *Radioactive gels.* Radioactive proteins can be detected in gels by autoradiography or fluorography (Laskey and Mills, 1975). Autoradiography is direct exposure of the gel to X-ray film. Fluorography, exposure of the gel after incorporation of a fluor, is the method used most frequently for detection of radioactive proteins because of its sensitivity. Extremely small quantities of protein can be localized and quantitated in any of the types of gels mentioned above using this method.

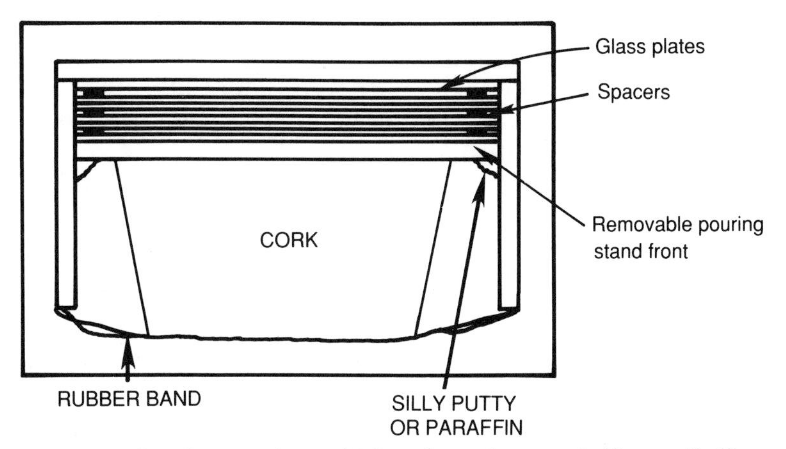

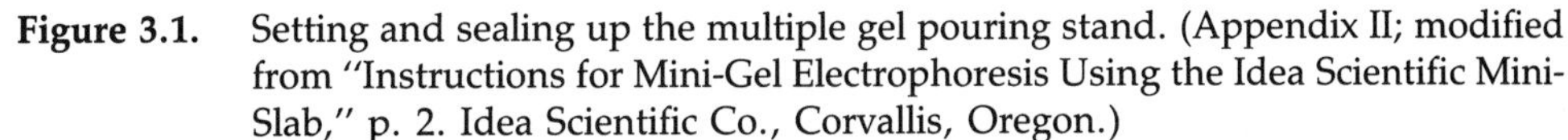

Figure 3.1. Setting and sealing up the multiple gel pouring stand. (Appendix II; modified from "Instructions for Mini-Gel Electrophoresis Using the Idea Scientific Mini-Slab," p. 2. Idea Scientific Co., Corvallis, Oregon.)

This experiment describes a protocol for electrophoresis in a 12.5% acrylamide gel with a 6% acrylamide stacking gel using the discontinuous buffer system originally described by Laemmli (1970) and for the staining, drying, and autoradiography of this gel. The procedure utilizes a microslab gel apparatus similar to the one described by Matsudaira and Burgess (1978) because electrophoresis in this system is extremely rapid and sensitive.

WARNING **Unpolymerized acrylamide is a toxin of the central nervous system! Always wear gloves when handling acrylamide or making acrylamide gels, and never pipet acrylamide solutions by mouth! Even after polymerization, a small amount of acrylamide monomer remains in the gel.**

Protocols

Procedure for Teaching Assistants

Teaching assistants should prepare this the day before use.

Preparation of the Gel Mold

1. Select 10, $3\frac{1}{4} \times 4$ inch glass plates and carefully clean them with warm soapy water. Rinse carefully in distilled water, and dry with Kimwipes or by standing them in a test tube rack with pins and allowing them to air dry.
2. Select 10, 0.8-mm-thick spacers. Form a series of five gel "sandwiches" in the multiple gel pouring stand with two side spacers in each sandwich (see Figure 3.1 above). Place

continued on next page

pouring stand front on the sandwiches and put a cork stopper against the front. Hold this arrangement in place with a weak rubber band (a strong rubber band will bow the plates).

3. Seal the front of the pouring stand in place with melted paraffin. Leaks will usually occur at the bottom corners, so seal carefully there.

A. Preparation of the 12.5% Acrylamide Separating Gel

1. In a 125-ml side-arm flask, mix 40 ml of separating gel solution as follows:

16.7 ml acrylamide stock (30:0.8)	12.5% acrylamide
5 ml running gel buffer (8×)	0.38 *M* Tris–HCl (pH 8.8)
0.2 ml 20% SDS	0.1% SDS
18.1 ml H_2O	
40.0 ml	

Stopper the flask and apply a vacuum for 3 to 5 minutes. Swirl the flask gently a few times.

Add 300 μl 10% ammonium persulfate (freshly prepared)
20 μl TEMED

Swirl the flask gently to mix. Be careful not to generate bubbles.

2. Pipet the solution into the gel sandwiches to a level about 2 cm below the top of the glass plates.
3. Overlay the gels with water-saturated isobutanol. Layer the isobutanol solution gently, using a Pasteur pipet.
4. A sharp butanol–gel interface should appear after 15 to 30 minutes under the isobutanol layer when the gel is polymerized.
5. Store gels for at least 2 hours. Gels can be kept for several days in a sealed plastic bag containing wet paper toweling.
6. Pour off the isobutanol. Rinse the surface of the gel several times with distilled water.

B. Preparation of the Stacking Gel

1. Pour off the liquid from the surface of the gel. Remove the polymerized gels from the casting stand. Try not to remove

the spacers. Using a razor blade, pry the individual gels apart and remove the polymerized acrylamide from the outsides of the plates with a scraping motion of the blade.

2. Clamp the side and bottom edges of the gel with small binder clips (two to three per side) and then seal the side edges with molten 2% agarose using a Pasteur pipet. This will allow pouring of the stacking gel.
3. In a 50-ml side-arm flask, mix 10.05 ml of stacking gel solution as follows:

2.0 ml acrylamide stock (30 : 0.8)	6% acrylamide
2.5 ml stacking gel buffer (4×)	0.12 *M* Tris–HCl (pH 6.8)
50 μl 20% SDS	0.1% SDS
5.5 ml H_2O	
10.05 ml	

Evacuate as before.

Add 75 μl 10% ammonium persulfate
5 μl TEMED

Swirl the flask gently to mix. Be careful not to generate bubbles.

4. Add about 2 to 3 ml of stacking gel solution to the sandwich to rinse the surface of the gel, then pour off the liquid.
5. Fill the sandwich to the top with the gel mixture.
6. Insert the comb into the gel, taking care not to allow air bubbles to collect under the "teeth." The comb should extend into the gel far enough to give a stacking gel of about 1 cm, as diagrammed in Figure 3.2.
7. Allow the gel to polymerize for about 30 minutes. When the process is complete you should be able to see Schlieren lines around the teeth of the gel combs. The gel can be loaded immediately after the stacking gel has polymerized.

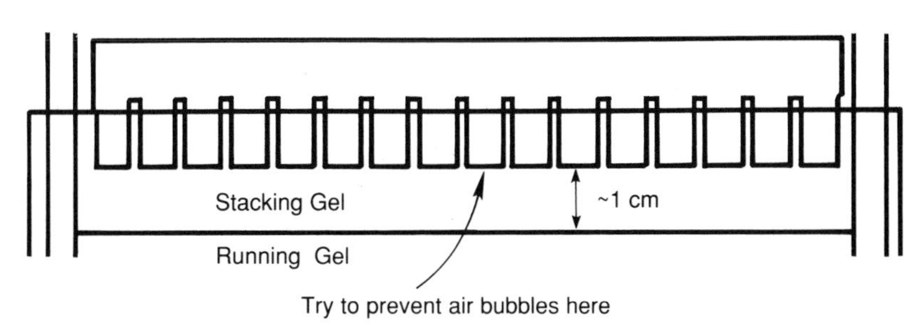

Figure 3.2. Correct positioning of the well-forming comb in an acrylamide gel. (Modified from "Instructions for Mini-Gel Electrophoresis Using the Idea Scientific Mini-Slab," p. 4. Idea Scientific Co., Corvallis, Oregon.)

C. Electrophoresis

1. Select one gel and insert it into the slot in the upper reservoir of the minigel apparatus. Clamp the gel in place with two small binder clamps. The slots formed by the gel comb should extend into the upper reservoir and be completely visible when viewed from the side (see Figure 3.3).
2. Seal the gel in place using 2% agarose. A Pasteur pipet works well for this procedure. Be sure to seal carefully around the entire gel as shown in Figure 3.3. Allow at least 15 to 20 minutes for the agarose to harden before adding buffer to the upper reservoir. During this time begin preparing the samples for electrophoresis (see next section for instructions).
3. Carefully remove the gel comb. Fill the upper buffer chamber with reservoir buffer. Using a Pasteur pipet, rinse the wells formed by the gel comb with reservoir buffer. The apparatus is now ready for the samples. Load the samples (as soon as possible after boiling) using a microliter syringe. The sample volume should be in the range of 5 to 30 μl and should contain no more than about 30 μg of protein. See the next section for details on appropriate loadings for chloroplast samples.

   Insert the syringe needle into the desired slot and **slowly** expel the contents of the syringe into the slot. Rinse the syringe after each sample (the upper reservoir buffer is convenient for this), and repeat until all samples have been loaded. Put the lid on the unit and connect to a power supply. The cathode (black lead) should be connected to the upper buffer chamber.

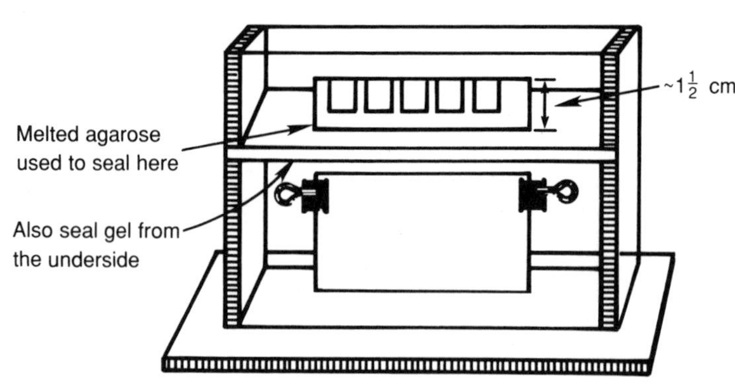

Figure 3.3. Sealing the acrylamide gel into the gel apparatus. (Modified from "Instructions for Mini-Gel Electrophoresis Using the Idea Scientific Mini-Slab," p. 6. Idea Scientific Co., Corvallis, Oregon.)

4. Set the power supply to constant wattage and adjust to 2.5 W/gel. Initial voltage and current settings should be about 100 V and 25 mA. The bromphenol blue tracking dye should be electrophoresed to about 0.5 mm from the bottom of the gel (approximately 90 to 120 minutes).
5. When the dye reaches the bottom of the slab gel, turn the power supply off and disconnect the power cables from the gel. **It is important to disconnect leads from power supply so that no one is accidentally shocked!!**

D. Sample Preparation for SDS–Gel Electrophoresis

1. In General

For this type of gel, proteins should be in a low ionic strength buffer (<100 m*M* Tris, etc.) in water or as a pellet. When possible, the protein concentration should be estimated by the Lowry assay and should be in the range of 0.5–10 mg/ml. Protein pellets can be resuspended directly in SDS sample buffer. This is particularly convenient for the thylakoid membrane fractions.

2. For Chloroplast Samples

Stroma: Fifty micrograms chlorophyll (equivalent)/200 μl, from Experiment 3A, step 17. Mix 200 μl of stroma directly with 200 μl of 2× SDS sample buffer in a microfuge tube. This will give the equivalent of 50 μg chlorophyll/400 μl. Heat the sample in a boiling water bath for 1 minute (be certain the tube is capped tightly). Load 32 μl; this will give the equivalent of proteins derived from chloroplasts containing 4 μg chlorophyll.

Thylakoids: Resuspend one of the pellets in 1× SDS sample buffer to give a concentration of 50 μg chlorophyll/400 μl sample buffer. Note that the membranes will have to be vortexed vigorously to be resuspended. Hold the membranes on ice. Remove an aliquot of 32 μl to a fresh microfuge tube. Cap the tube tightly and incubate in a boiling water bath for *no more than 20 seconds*. Immediately chill the heated sample on ice. Load 32 μl of heated thylakoids, and in a separate lane load 32 μl of nonheated thylakoids. This will give the equivalent of 4 μg chlorophyll in each sample.

Whole Chloroplasts: Take 200 μl of the lysed chloroplast suspension and mix directly with 200 μl of 2× SDS sample buffer in a microfuge tube. Cap tightly, boil for 1 minute, and load 16 μl on the gel. This will give the equivalent of 2 μg of chlorophyll.

RuBisCo Standard: Boil 1 minute and load 20 μl. This gives about 500 ng of RuBisCo protein. The subunits run as polypeptides of roughly 50,000 and 12,000 to 14,000 molecular weight.

Molecular Weight Standards: Boil 1 minute and load 10 μl. This gives stainable bands for the following polypeptides:

Polypeptide	Molecular weight
Phosphorylase B	92,500
Bovine serum albumin	66,200
Ovalbumin	45,000
Carbonic anhydrase	31,000
Soybean trypsin inhibitor	21,500
Lysozyme	14,400

Crude Chloroplasts: Take an aliquot corresponding to 2 μg of chlorophyll. Mix with an equal volume of 2× SDS sample buffer, boil 1 minute, and load the entire sample.

Note: Save any remaining samples at −20 or −70°C in case any problems are encountered with the gel. The samples can be used again.

E. Staining and Destaining

1. Discard the reservoir buffer in the upper chamber. Remove the gel sandwich from the apparatus and remove the spacers from the sides. Using a razor blade as a spatula, **gently** pry the plates apart. Be careful that the gel is adhering to one plate only or it will tear.
2. Add about 100 ml of stain solution to a plastic staining box. Put the gel into the stain solution by inverting the gel over the box containing the stain; if the gel adheres to the plate, tease one corner of the gel off the plate with a spatula; touch this edge to the surface of the stain solution and the gel should peel away from the plate without further assistance.
3. Cover the staining box and gently shake the gel for 30 minutes to overnight.
4. Pour off the stain and cover the gel with destain solution. Gently shake the gel for 30 minutes.

5. Repeat step 4 until the gel is destained to the appropriate level. A wadded Kimwipe included in the box along the edge will absorb stain and permit faster destaining of the gel.
6. If the gel contains radioactively labeled proteins, fluorography can be carried out after Coomassie blue staining by
   a. soaking the gel for 30 minutes in H_2O
   b. soaking the gel twice for 30 minutes each time in Autofluor or Fluoro-Hance
   c. drying the gel by placing it on two sheets of Whatman 3MM filter paper, covering it with plastic wrap, and placing it on a gel drier set at 60°C for 1 hour; when the gel is dry, peel off the plastic wrap and expose the gel to X-ray film at −70°C; a gel containing 10^5 cpm of ^3H- or ^{35}S-labeled proteins per lane will take 24 to 72 hours for a reasonable exposure

Materials Provided

Solution	Preparation
Acrylamide stock (30 : 0.8)	
30% (w/v) acrylamide	58.4 g acrylamide
0.8% (w/v) bisacrylamide	1.6 g methylene-bisacrylamide
	adjust final volume to 200 ml with distilled water
	filter through 0.22-μm nitrocellulose
	store at 4°C in the dark
10% ammonium persulfate	0.5 g ammonium persulfate
	adjust final volume to 5 ml with distilled water
	store at 4°C for no longer than 2 weeks
Running gel buffer (8×)	
3 *M* Tris–HCl (pH 8.8)	181.7 g Tris base
	add 250 ml water
	adjust pH to 8.8 with conc. HCl
	adjust final volume to 500 ml with distilled water
	filter through 0.22-μm nitrocellulose
	store at room temperature

Solution	Preparation
Stacking gel buffer (4×) 0.5 *M* Tris–HCl (pH 6.8)	15.1 g Tris base add 150 ml water adjust pH to 6.8 with conc. HCl ——— adjust final volume to 250 ml with distilled water filter through 0.22-μm nitrocellulose store at room temperature
20% (w/v) SDS	40 g SDS ——— adjust final volume to 200 ml with distilled water filter through 0.22-μm nitrocellulose store at room temperature
Reservoir buffer 25 m*M* Tris 0.192 *M* glycine 0.1% (w/v) SDS	12.0 g Tris base 57.6 g glycine 20 ml 20% SDS ——— adjust final volume to 4 liters with distilled water pH of this solution (~8.3) should not require adjustment store at room temperature
2× sample loading buffer 0.12 *M* Tris 4% (w/v) SDS 10% 2-mercaptoethanol 20% (v/v) glycerol 2 mg/ml bromphenol blue	0.3 g Tris 4 ml 20% SDS 2 ml 2-mercaptoethanol 4 ml glycerol 40 mg bromphenol blue 10 ml water ——— store at −20°C in 1-ml aliquots
Stain solution 0.125% (w/v) Coomassie blue 50% (v/v) methanol 10% (v/v) acetic acid	1.25 g Coomassie blue R250 500 ml methanol 100 ml glacial acetic acid ——— adjust final volume to 1 liter with distilled water filter through Whatman #1 filter

Solution	Preparation
Destain solution 10% (v/v) methanol 7% (v/v) acetic acid	1 liter methanol 0.7 liter glacial acetic acid ——— adjust final volume to 10 liters with distilled water
Water-saturated isobutanol	100 ml isobutanol 100 ml distilled water ——— mix together and shake well; two phases should form, butanol on the top and water on the bottom store in a tightly capped bottle
2% (w/v) agarose	2 g agarose 100 ml distilled water ——— dissolve agarose by boiling for several minutes distribute into 10- or 20-ml aliquots cap tightly and store at room temperature
Molecular weight standards	50 μl low molecular weight SDS–PAGE standards (Bio-Rad #161-0304) 500 μl 2× sample loading buffer 450 μl distilled water ——— divide into 100-μl aliquots store at −20°C
RuBisCo standard	1 U RuBisCo (Sigma #R8000), about 300–1000 μg ——— adjust volume to give 0.05 μg/μl in 1× sample loading buffer store at −20°C in 100- to 500-μl aliquots

References

Hames, B. D., and Rickwood, D., eds. (1981). "Gel Electrophoresis of Proteins." IRL Press Limited, Oxford.

Hedrick, J. L., and Smith, A. J. (1968). Size and charge isomer separation and estimation of molecular weights of proteins by disc gel electrophoresis. *Arch. Biochem. Biophys.* **126,** 155–164.

Hurkman, W. J., and Tanaka, C. K. (1986). Solubilization of plant membrane proteins for analysis by two-dimensional gel electrophoresis. *Plant Physiol.* **81,** 802–806.

Laemmli, U. K. (1970). Cleavage of structural proteins during the assembly of the head of bacteriophage T4. *Nature (London)* **227,** 680–685.

Laskey, R. A., and Mills, A. D. (1975). Quantitative film detection of ^{3}H and ^{14}C in polyacrylamide gels by fluorography. *Eur. J. Biochem.* **56,** 335–341.

Matsudaira, P. T., and Burgess, D. R. (1978). SDS microslab linear gradient polyacrylamide gel electrophoresis. *Anal. Biochem.* **87,** 387–396.

Neville, D. M. (1971). Molecular weight determination of protein–dodecyl sulfate complexes by gel electrophoresis in a discontinuous buffer system. *J. Biol. Chem.* **246,** 6328–6334.

O'Farrell, P. H. (1975). High-resolution two-dimensional electrophoresis of proteins. *J. Biol. Chem.* **250,** 4007–4021.

Piccioni, R., Bellemare, G., and Chua, N.-H. (1982). Methods of polyacrylamide gel electrophoresis in the analysis and preparation of plant polypeptides. *In* "Methods in Chloroplast Molecular Biology" (M. Edelman, R. B. Hallick, and N.-H. Chua, eds.), pp. 985–1041. Elsevier, Amsterdam.

Schleif, R. F., and Wensink, P. C. (1981). "Practical Methods in Molecular Biology," pp. 79–88. Springer-Verlag, Berlin and New York.

Tanford, C., Nozaki, Y., Reynolds, J. A., and Makino, S. (1974). Molecular characterization of proteins in detergent solutions. *Biochemistry* **13,** 2369–2376.

Weber, K., and Osborn, M. (1969). The reliability of molecular weight determinations by dodecyl sulfate–polyacrylamide gel electrophoresis. *J. Biol. Chem.* **244,** 4406–4412.

EXPERIMENT

4 Isolation of Chloroplast DNA

Introduction

Plant cells require coordination between three distinct, but interdependent genetic systems during their course of development. In addition to the genetic information housed in the nucleus, plant cells contain DNA in their chloroplasts and mitochondria. These organellar DNAs resemble both bacterial and eukaryotic nuclear DNAs in their organization (e.g., they do not have the nucleoprotein organization characteristic of nuclear DNA, but several genes have introns), and they encode some, but not all, of the information necessary to ensure growth and replication of chloroplasts and mitochondria. Unlike mitochondrial DNA, which normally accounts for only about 1% of the DNA in a plant cell, chloroplast DNA can represent a substantial portion (~30%) of the DNA in a mature leaf cell (Palmer, 1985). Chloroplast DNA exists as closed, circular molecules of about 150 (±30) kb in most plants. A common feature of chloroplast DNA is a large, inverted repeat sequence (~10–25 kb in length) separating two single copy regions (~80 kb and ~20 kb) (Kolodner and Tewari, 1979; Palmer, 1985). Despite the size differences, the gene content and overall organization of chloroplast DNAs from widely divergent species is remarkably similar. Within any single plant, chloroplast DNA appears to consist of two equimolar populations of molecules that differ only in the relative orientation of their single copy sequences (Palmer, 1983). However, individual genes can be rearranged in their relative positions on the chloroplast DNA maps of different plant species (Palmer and Thompson, 1981).

Recently, the complete nucleotide sequences of chloroplast DNA from a liverwort (Ohyama *et al.*, 1986) and an angiosperm (Shinozaki *et al.*, 1986) have been determined. Figure 4.1 shows a genetic map of tobacco chloroplast DNA. Sequence analysis

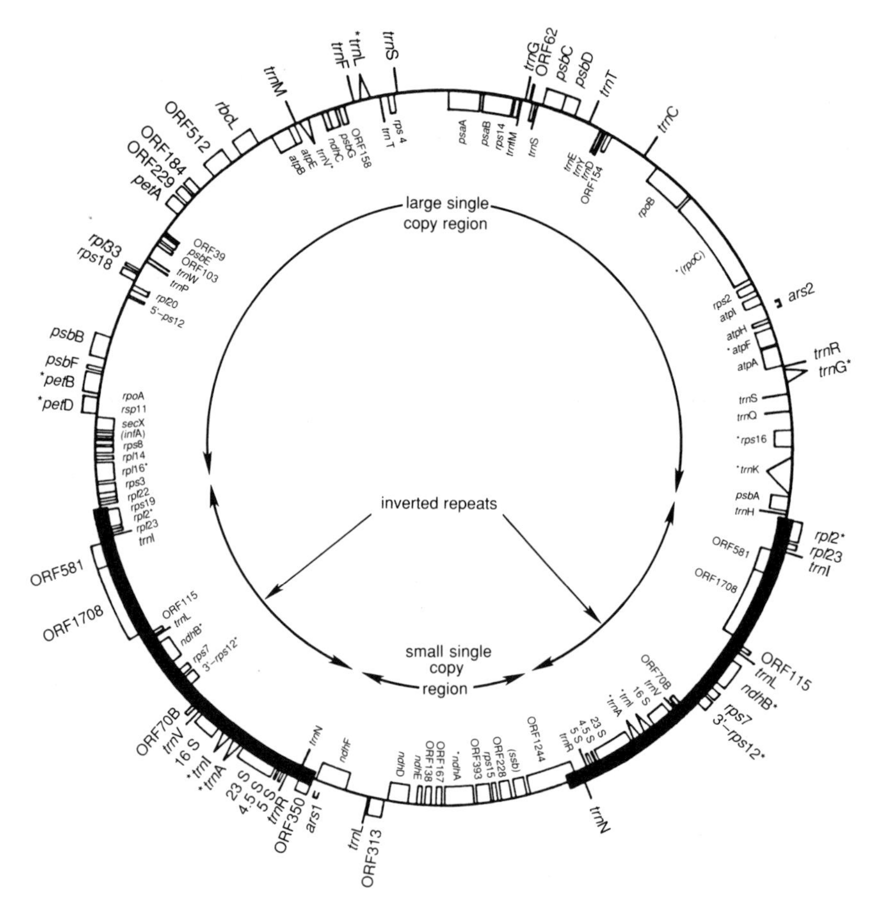

Figure 4.1. Genetic map of the 155,844-bp circular chloroplast DNA of tobacco (*Nicotiana tabacum* var. "Bright Yellow 4"). The large and small single copy regions and the inverted repeat regions are indicated. Some of the abbreviations used to designate different genes include *trn* (tRNA); *psa* (photosystem I); *psb* (photosystem II); *atp* (coupling factor ATPase); and ORF (unidentified open reading frame). [For complete details, see Shinozaki *et al.* (1986).]

predicts this DNA to contain over 120 genes, including those encoding 4 different rRNAs, 30 tRNAs, 39 polypeptides identified in mapping studies or by sequence similarities with other organisms, and 11 other predicted polypeptide-coding genes (open reading frames or ORFs).

In order to adequately separate the chloroplast DNA from the nuclear DNA of an organism, the two DNA populations must differ enough in their buoyant densities so that they can be separated on isopycnic CsCl density gradients. Alternatively, one can enrich for chloroplast DNA by first isolating intact chloroplasts and then by isolating plastid DNA. The large size of the circular chloroplast genome (120–190 kb) and its fragility necessitate that the utmost care be used during the isolation so that nicks are not introduced into the chloroplast genome (Herrmann, 1982). In the DNA isolation procedure outlined below, chloroplasts are purified from a 30:45:60% sucrose step gradient (Palmer, 1982). Intact chloroplasts purified in this way are used because it is important to remove fragmented organelles and nuclei before attempting to isolate the chloroplast DNA. Nuclear DNA is usually the major contaminant in chloroplast DNA preparations, and fragmented chloroplasts may contain partially degraded DNA. Although the sucrose step gradients frequently disrupt physiological functions occurring in the chloroplast, this procedure has little effect on the integrity of the chloroplast DNA. In many cases, intact and fragmented chloroplasts can be differentiated easily with a phase contrast microscope: the intact chloroplasts are cupped and have distinct outlines; the fragmented chloroplasts are flattened and have irregular shapes. Alternatively, intact chloroplasts purified by centrifugation through Percoll gradients may also be used.

Because the chloroplast DNA is tightly associated with chloroplast membranes, the lysis procedure is one of the most critical steps in this procedure. For this reason, after the intact chloroplast fraction is obtained, the organelles are lysed and deproteinated by the addition of proteinase K and sarkosyl (a detergent). Proteinase K (named for its keratinolytic activity) is an extremely stable fungal protease that is also active at high temperatures (65°C) and in the presence of urea, detergents (SDS and sarkosyl), and EDTA. Its primary usefulness, however, is that it rapidly inactivates endogenous RNases and DNases present in the tissue of interest. After deproteinization, CsCl and ethidium bromide are added to the chloroplast extract and the DNAs are separated by equilibrium centrifugation. The chloroplast DNA band is removed from the gradient, extracted with isopropanol to remove ethidium bromide, and precipitated with ethanol. Subsequent steps in this experiment involve restriction analysis of the purified chloroplast DNA and DNA Southern analysis to map the genes in the chloroplast genome for the large subunit of RuBisCo (*rbcL*) (McIntosh *et al.*, 1980) and the ATPase β-subunit (*atpB*) (Krebbers *et al.*, 1982).

Protocols

A. Isolation of Chloroplast DNA

If glassware is used when working with small amounts of nucleic acid, it should be siliconized ("silanized") to minimize loss of nucleic acid due to binding to the glass. Alternatively, "plasticware" (polypropylene and polystyrene) can be used without silanization.

1. Bring chloroplast sample up to 2.7 ml with wash buffer (for composition, see Materials Provided section).
2. Add 0.3 ml of 10 mg/ml proteinase K solution (for composition, see Materials Provided section). Mix gently by inversion; incubate at room temperature for 15 minutes.
3. Add 0.3 ml of 20% sarkosyl. Mix by gentle inversions. Once sarkosyl has been added, the solution should be handled *very gently* in subsequent steps to minimize shearing of the high molecular weight DNA. Incubate the mixture at room temperature for 30–45 minutes, gently inverting the solution about 30–50 times during this period.
4. Add 1.0 ml of a 0.672 g/ml CsCl stock solution. Mix with gentle inversions (about 30–50 times) and allow the solution to stand 1 hour (or overnight) at 4°C.
5. Pellet membranes and starch particles by centrifuging sample for 20 minutes at 12,000 *g* (10,000 rpm in a Beckman JA17 rotor).
6. Carefully decant DNA into a 10-ml graduated cylinder or test tube. Adjust volume to 4.8 ml per gradient with Tris–EDTA buffer. For each gradient, add 6.7 ml of the 1.24 g/ml CsCl stock solution and 0.3 ml of the 10 mg/ml ethidium bromide (EtBr) solution. Invert slowly to mix solutions. After EtBr has been added to the DNA, the DNA–EtBr solution should be handled in subdued light to minimize light-catalyzed nicking of the DNA.
7. To assemble CsCl–EtBr gradients
   a. transfer the 11.8 ml of DNA–CsCl solution to a polycarbonate ultracentrifuge tube
   b. weigh the ultracentrifuge tubes and balance with CsCl "balance" solution
   c. attach the cap assemblies to the tubes and fill tubes completely with CsCl balance solution; balance tubes again and screw the cap assembly closed; a small air bubble in the centrifuge tube will not cause the tubes to collapse

   Note: "Quik-Seal" test tubes can be used instead of polycarbonate ultracentrifuge tubes with cap assemblies. When using Quik-Seal tubes, a metal adapter should be placed over the tube during ultracentrifugation, otherwise the tube will collapse.

8. Centrifuge the tubes in the Beckman 50 Ti rotor at 40,000 rpm for 40 hours, or in the 70.1 Ti rotor at 53,000 rpm for 16 to 20 hours, at 20°C.*

*Student Note ▫ Low temperature (4°C) will cause the CsCl to precipitate out of solution and change the CsCl density in the gradient. Consequently the DNA will band too low in the centrifuge tube.

9. The relaxed and supercoiled DNA bands present in the gradient are visualized with long-wavelength UV light, which minimizes nicking of the DNA.

a. Secure centrifuge tube to a ringstand with a clamp and remove the set screw. If using Quik-Seal tubes, puncture top of the test tube to prevent a vacuum from forming in step d below

b. Visualize DNA bands with a 360 nm UV source. Usually the chloroplast DNA is present in the relaxed and linear forms as is depicted in Figure 4.2

c. If using polycarbonate ultracentrifuge tubes, expel air from a sterile Pasteur pipet, gently move it down through the gradient, and position it just under the DNA band. Withdraw a minimal volume of DNA and transfer it to a plastic 15- or 50-ml screw-cap test tube

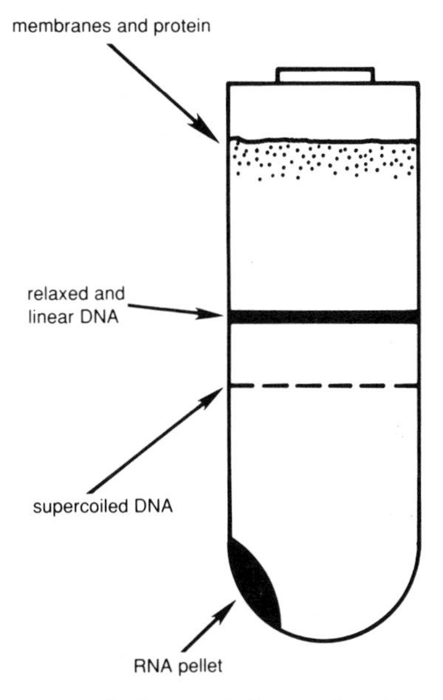

Figure 4.2. Appearance of UV-illuminated ultracentrifuge tube after equilibrium sedimentation of chloroplast DNA-containing solution. The DNA zones will have an orange/red fluorescent appearance.

d. If using Quik-Seal tubes, lightly smear acetone over the area of the tube where the DNA has banded. Wipe dry, then put a piece of sticky medical adhesive tape there. Lightly smear the adhesive tape with acetone and wipe dry. Carefully puncture the tape with a needle and syringe (18- or 16-gauge large-bore needle). *Do not* position your hand on the test tube wall opposite the point of needle insertion. Withdraw a minimal volume of DNA using a 5-ml syringe. Very gently expel the DNA from the syringe into a plastic 15- or 50-ml screw-cap test tube. If relaxed and supercoiled DNAs are being collected separately, carefully withdraw the needle and seal the hole with a second piece of adhesive tape

10. Note the original volume of the CsCl-banded DNA solution (usually 1 to 2 ml). Gently extract the DNA with an equal volume of buffer-saturated isopropanol (isopropanol is miscible with water but not with a concentrated CsCl solution). Gently invert the test tube several times and then allow the phases to separate by gravity. The ethidium bromide should partition into the upper (isopropanol) phase, which can be removed with a Pasteur pipet and discarded in the appropriate waste container (EtBr waste should be chemically inactivated before disposal). The aqueous phase should be reextracted with isopropanol until the aqueous phase is no longer pink (usually three to six extractions altogether). After the final extraction, adjust the volume of the DNA–CsCl solution to its original level using sterile water (the cloudiness of the solution should clear up). Transfer the aqueous phase to a (siliconized) 15-ml Corex test tube, being careful not to transfer any of the interface material.

11. Estimate the DNA volume and precipitate the DNA by adding

1 volume of TE.1 buffer (to dilute out the CsCl)
2 volumes ethanol (final conc. = 50% EtOH)

12. Chill the solution at −20°C for at least 4 hours and preferably overnight. (Do not chill at −70°C because the CsCl will precipitate out.)

13. Pellet the DNA by centrifugation at 12,000 *g* (10,000 rpm in a Beckman JA17 rotor) for 30 minutes. Slowly decant the supernatant and let the Corex tube drain upside down in a test tube rack. The pellet can be left to air dry for 15–60 minutes in this fashion. Save the pellet. Dissolve the DNA in 400 μl sterile TE.1 buffer. Allow the DNA to resuspend for at least 4 hours at 4°C (preferably overnight). Store the DNA at 4°C in microfuge tubes. (High molecular weight DNAs take a long time to resuspend so it's best to store them overnight before going on. Vacuum drying accentuates the problems associated with resuspending high molecular weight DNA.)

14. A second ethanol precipitation of the chloroplast DNA is usually required to fully remove residual CsCl. Residual CsCl can interfere with subsequent restriction enzyme digestions. The second ethanol precipitation can be performed in a 1.5-ml microfuge test tube with 0.1 volume of 3 *M* sodium acetate and 2.0 volumes of 100% ethanol.

15. Resuspend final DNA pellet in 100 μl of TE.1. You will have too little DNA to measure with the spectrophotometer. Instead, the amount of DNA can be estimated on a gel, relative to a known amount of *Hin*dIII-cut λ DNA.

B. Restriction Digestion of Chloroplast DNA

1. Set up 40 μl restriction digestions as you did in Experiment 1:

4 μl 10× *Eco*RI buffer
20 μl chloroplast DNA
15 μl sterile water
1 μl 5 U/μl *Eco*RI

40 μl

4 μl 10× *Pst*I buffer
20 μl chloroplast DNA
15 μl sterile water
1 μl 5 U/μl *Pst*I buffer

40 μl

4 μl 10× *Hin*dIII buffer
20 μl chloroplast DNA
15 μl sterile water
1 μl 5 U/μl *Hin*dIII

40 μl

4 μl 10× *Bam*HI buffer
20 μl chloroplast DNA
15 μl sterile water
1 μl 5 U/μl *Bam*HI

40 μl

2. Incubate restriction digestions for 45 minutes at 37°C. At the end of this period, add 8 μl 10 *M* urea loading dye. Load 24 μl of sample per lane on gel. Don't forget to load the ^{32}P-labeled *Hin*dIII-cut λ DNA standard.
3. Pour a 0.8% agarose gel as follows*:
   a. dissolve 1.2 g agarose in 150 ml 1× agarose gel buffer
   b. heat in microwave or on hot plate until completely dissolved; cool slightly; pour into gel plate; adjust 13-well comb
   c. load two sets of samples onto your gel (because we will probe the Southerns with two different probes); the order should be *Hin*d-cut λ DNA, *Eco*RI DNA, *Pst*I DNA, *Hin*dIII DNA, *Bam*HI DNA, two spaces, *Hin*d-cut λ DNA, *Eco*RI DNA, etc.

*Student Note

□ We use lower percentage agarose gels for mapping the chloroplast DNA because these gels separate the higher molecular weight fragments better.

4. Run gel either at 20 V overnight or at 100 V for 4 hours until bromphenol blue dye reaches the bottom of the gel.
5. Stain the agarose gel with 0.5 μg/ml EtBr in water for 30 minutes. Photograph gel.

C. Genomic DNA Southern Transfer

1. Soak gel in 250 ml (enough to cover gel) 0.25 *M* HCl† for 15 minutes with occasional swirling; drain; repeat a second time with 0.25 *M* HCl for 15 minutes. Acid treatment will depurinate DNA.

†Student Notes

□ Be very careful about the times for the HCl and later washes. Too long an HCl treatment will cleave the DNA into tiny fragments which cannot stick to nitrocellulose; long NaOH and Tris washes will allow the smaller DNAs to diffuse out of the gel.

□ The layers of paper and towels over the slab gel must not contact the wet paper below because this will short circuit the liquid passing through the gel, nitrocellulose, and into the paper towels.

□ The nitrocellulose filters can be marked with India ink or a permanent marker pen before putting them on top of the agarose gel.

□ If the filters are baked too long or at too high a temperature they become too brittle to work with—**be careful!**

2. Rinse gel once briefly with distilled water.
3. Soak gel in 250 ml 0.5 *M* NaOH, 1 *M* NaCl for 15 minutes; drain; repeat a second time with 0.5 *M* NaOH, 1 *M* NaCl for 15 minutes; drain. Base treatment will cleave DNA at apurinic residues and denature the DNA fragments prior to transfer.
4. Soak gel in 250 ml 0.5 *M* Tris–HCl (pH 7.5), 3 *M* NaCl for 5 minutes; drain. Repeat with 0.5 *M* Tris–HCl (pH 7.5), 3 *M* NaCl for 15 minutes; drain.
5. Set the gel up for transfer to nitrocellulose paper as is shown in Figure 4.3.
6. Allow DNA to transfer to nitrocellulose filter for 3 hours or overnight at 25°C. The agarose gel should be paper thin after 3 hours. Cut the filter in half since you will hybridize each half with a different probe. Place filters on a paper towel and dry for 1 hour at 65–70°C (*no higher*) in a vacuum oven.

Southerns set up this way should be allowed to transfer for 2 hours or overnight at room temperature and then bake for 1 hour at 65–70°C.

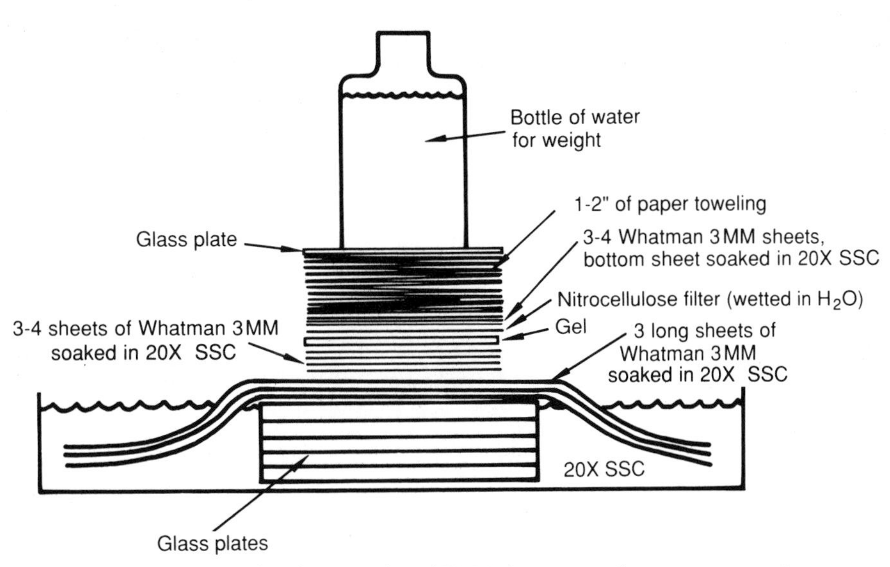

Figure 4.3. Standard arrangement for the transfer of DNA fragments from agarose gels to nitrocellulose filters (Southern, 1975). [Figure modified from Schleif and Wensink (1981), p. 153.]

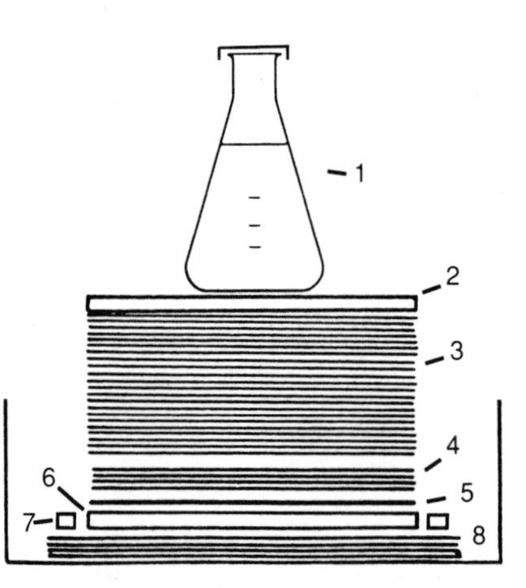

Figure 4.4. Set-up for simplified Southern blots.*
1. Flask or bottle with water
2. Glass plate
3. Paper toweling, about 1½–2 inches, cut to fit over 3MM filters
4. Whatman 3MM filters, three or four, cut to fit over nitrocellulose
5. Nitrocellulose filter
6. Agarose gel
7. Plastic strips or four edges of agarose gel (to prevent contact between the reservoir and wick)
8. Whatman 3MM filters, three or four, cut to fit bottom of dish and saturated with 20× SSC

*Student Note ▫ There is a simpler Southern set-up for gels (Figure 4.4) that contain plasmid DNA fragments rather than genomic DNA fragments. Because plasmid DNA fragments are generally in much higher concentrations than genomic DNA fragments, plasmid DNAs can be transferred from the gel to nitrocellulose with minimal amounts of 20× SSC buffer.

D. Nick Translation

This procedure is the same as in Experiment 2. You will be provided with 0.5 μg/μl of a 593-bp *Bgl*I·*Pst*I fragment from pZmc702, a subclone of the maize ATPase β-subunit gene, and a 567-bp *Pst*I fragment from pZmc701, a subclone of the maize large subunit RuBisCo gene.

1. Set up the nick-translation procedure as follows:

10 μl DNA (0.2–2.0 μg purified DNA fragment)
10 μl 10× nick-translation buffer [500 m*M* Tris (pH 7.8), 50 m*M* $MgCl_2$, 1 mg/ml BSA, 10 μ*M* 2-mercapto-ethanol]
2 μl of 20 m*M* dCTP, dGTP, dTTP stock
2 μl [^{32}P]dATP (10 μCi/μl, 3000 Ci/mmol)

74 μl H_2O
1 μl 1/10,000 dilution of 1 mg/ml stock DNase (dilute immediately before use)
1 μl DNA polymerase I holoenzyme

100 μl reaction

Incubate for 1 hour at 15°C in an ice bucket water bath behind isotope shield.

Add 20 μl nick-translation stop buffer.

2. Set up a 4-ml G-100 Sephadex column to separate incorporated and unincorporated ^{32}P. Stopper a plastic 5-ml pipet with siliconized glass wool. Fill pipet with TE buffer and add G-100 Sephadex until column bed column = 4 ml. Keep buffer on top of the Sephadex so that column doesn't run dry. Layer nick-translation reaction onto top of column. After it has adsorbed to the column, start eluting with sterile TE buffer. Collect the first blue peak that comes off the column—this excluded peak contains the incorporated radioactive nucleotides. Dispose of the unincorporated nucleotide (the second blue band) and the column in the radioactive waste area.
3. Count 1 μl of the probe on a paper circle in a scintillation vial without scintillant on ^3H channel. You should get 5,000–20,000 cpm/μl when purified DNA fragment is used for nick translation.

E. Hybridization of Southerns

1. Seal nitrocellulose filter into seal-a-meal bag; check to make sure that the seals are tight.
2. Mix prehybridization solution as follows:

   7.5 ml 10× SSC, 0.4% sarkosyl, 2× Denhardt's solution
   6.0 ml formamide (final conc. = 40% formamide)
   1.5 ml H_2O

   15.0 ml

3. Cut off one corner of bag. Add prehybridization solution to bag. Reseal. Incubate for 10 minutes on your benchtop. Squeeze out excess liquid. (Usually the prehybridization step is performed for 4 hours to overnight to minimize background hybridization.)
4. Add 500,000 cpm of each probe to a separate test tube containing 1 ml of sterile water. Boil the ^{32}P nick-translated DNA for 3 minutes in a boiling water bath to separate the DNA strands.

Mix the heated probe with the hybridization solution (outlined below) and add to the seal-a-meal bag containing the filters. Try to get rid of as many bubbles as possible before resealing the bag.

Set up each bag of filters with 15 ml of hybridization solution as follows:

7.5 ml	10× SSC, 0.4% sarkosyl, 2× Denhardt's solution
6.0 ml	formamide
0.5 ml	H_2O
1.0 ml	^{32}P nick-translated DNA in H_2O (use approximately 500,000 cpm/filter)
15.0 ml	

Hybridize filters for 12–20 hours (overnight) at 40°C. The relatively low hybridization temperature is due to the following: (1) the 37–38% GC content of the chloroplast genome and (2) the use of a heterologous probe for hybridization.

5. To wash off the unhybridized probe, cut a corner of bag, pour off hybridization solution, open bag up, and float filters in a Pyrex dish at 50°C in 2× SSC, 0.5% sarkosyl four times for 20 minutes each time. If possible, use a shaking water bath or a shaker–incubator. Dry the filters on a paper towel at room temperature. Mark the filters with ^{32}P-labeled ink. Put Saran wrap over the filters so they don't stick to the film. Set up an autoradiogram with XAR-5 film and expose in the −70°C freezer for 3 days (or until the next class period.)

Procedure for Teaching Assistants

Sucrose Gradient Method for the Isolation of Chloroplasts

1. Use 20 g of fresh leaf tissues/2 students; try to destarch material by storing plants in the dark 1–2 days before prep. If leaf material is removed from plants prior to prep, store in plastic bag with wet paper towels in the dark and in the refrigerator.

2. Remove young growing leaves from plants for the prep. Excise away nongreen and necrotic tissue. Rinse in cold water several times; rinse in 50 ml cold 10% Clorox 1–2 minutes; rinse extensively with cold water, being sure to rinse off bleach. Blot tissue dry with towels or use salad spinner. Cut into small pieces if necessary.

continued on next page

3. Repair to the cold room. Blend tissue in 5 volumes/gram fresh weight of extraction buffer (100 ml) in a Waring blender. Use two or three 5-second bursts on low speed.
4. Filter rapidly through six layers of cheese cloth and one layer of Miracloth (gravity filtration).
5. Pour filtrate into 250- or 500-ml centrifuge bottles; spin at 250 rpm in GSA, GS3, or HB4 for 15 minutes at 4°C.
6. Discard supernatant and add 8 ml of wash buffer (see protocol for chloroplast DNA isolation). Resuspend green pellet gently with fine camel's hair brush, swirling contents in centrifuge bottle. **Do not vortex.**
7. Layer resuspended organelles over a 30:45:60% sucrose step gradient. Top off gradient with wash buffer. Balance tubes. Centrifuge in the SW27 (Beckman) or AH637 (Sorvall) rotor at 25,000 rpm for 60 minutes at 4°C.
8. Use a wide-bore pipet or cut-off pipet tip to remove intact chloroplast band, the one at or closest to the 30:45% interphase. It should be very dark green.
9. Estimate volume of chloroplast band. Add 3 volumes cold wash buffer. Centrifuge at 4000 rpm for 10 minutes at 4°C in the SS34 rotor. Resuspend pellet in 9 ml wash buffer and hold on ice or in refrigerator until students need the chloroplasts.

Materials Provided

A. Isolation of Chloroplast DNA

Chloroplasts purified (by teaching assistant) from either 30:45:60% sucrose gradients or from Percoll gradients

Wash buffer	
350 m*M* sorbitol	6.4 g sorbitol
50 m*M* Tris–HCl (pH 8.0)	5.0 ml 1 *M* Tris–HCl (pH 8.0)
20 m*M* EDTA	10 ml 200 m*M* EDTA
	adjust final volume to 100 ml with distilled water
	autoclave

10 mg/ml proteinase K solution	100 mg proteinase K
	10 ml wash buffer
	store at −20°C

20% (w/v) sarkosyl	20 g sodium lauryl sarkosyl in 100 ml sterile water

Sterile 10-ml graduated cylinders (or test tubes)

Sterile plastic test tubes for high-speed centrifuge

Sterile, siliconized 15-ml Corex text tubes

0.672 g/ml CsCl stock solution	26.88 CsCl
	adjust final volume to 40 ml with 50 m*M* Tris–HCl (pH 8), 20 m*M* EDTA
1.24 g/ml CsCl stock solution	248 g CsCl
	adjust final volume to 200 ml with 50 m*M* Tris–HCl (pH 8), 20 m*M* EDTA
CsCl "balance" solution	19.0 ml wash buffer
	1.5 ml 20% sarkosyl
	5.0 ml 0.672 g/ml CsCl
	33.5 ml 1.24 g/ml CsCl
	49.0 ml
10 mg/ml ethidum bromide stock	100 mg ethidium bromide
	adjust final volume to 10 ml with sterile water
Tris–EDTA buffer	
50 m*M* Tris–HCl (pH 8.0)	6.06 g Tris base
20 m*M* EDTA	7.44 g $Na_2EDTA \cdot 2H_2O$
	adjust final volume to 1 liter with distilled water
	dissolve in 900 ml distilled water
	adjust pH to 8.0 with HCl
	autoclave
Buffer-saturated isopropanol	15 ml TE buffer
	10 g CsCl
	dissolve CsCl in TE buffer, then add 100 ml isopropanol; isopropanol should be buffer saturated for at least 24 hours

TE buffer	
10 m*M* Tris–HCl (pH 8.0)	1.21 g Tris base
1 m*M* EDTA	0.37 g $Na_2EDTA \cdot 2H_2O$
	dissolve in 900 ml distilled water adjust pH to 8.0 with HCl adjust final volume to 1 liter with distilled water autoclave
TE.1 buffer	
10 m*M* Tris–HCl (pH 8.0)	1.21 g Tris base
0.1 m*M* EDTA	0.04 g $Na_2EDTA \cdot 2H_2O$
	dissolve in 900 ml distilled water adjust pH to 8.0 with HCl adjust final volume to 1 liter with distilled water autoclave
2 *M* NaCl	11.7 g NaCl
	adjust final volume to 100 ml autoclave

B. Restriction Digestion of Chloroplast DNA and Genomic DNA Southern Transfers

10× *Eco*RI buffer
10× *Pst*I buffer
10× *Bam*HI buffer
Sterile water
10 *M* urea loading dye
^{32}P-Labeled *Hin*dIII-cut λ DNA standards

2× hybridization buffer	
10× SSC	88 g NaCl (1.5 *M*) 44 g sodium citrate (0.15 *M*)
0.4% sarkosyl	4 g sarkosyl
2× Denhardt's solution	
0.04% BSA	0.4 g BSA
0.04% polyvinylpyrrolidone	0.4 g polyvinylpyrrolidone
0.04% Ficoll	0.4 g Ficoll
	adjust final volume to 1 liter with distilled water

0.25 *M* HCl	40 ml conc. HCl
	adjust final volume to 2 liters with distilled water
0.5 *M* NaOH	40 g NaOH
1 *M* NaCl	116 g NaCl
	adjust final volume to 2 liters with distilled water
0.5 *M* Tris (pH 7.5)	121 g Tris base
3 *M* NaCl	350 g NaCl
	add 1.8 liters water
	add conc. HCl to pH 7.5
	adjust final volume to 2 liters with distilled water
20× SSC	
3 *M* NaCl	350 g NaCl
0.3 *M* sodium citrate	176 g sodium citrate
	adjust final volume to 2 liters with distilled water

References

Herrmann, R. G. (1982). The preparation of circular DNA from plastids. *In* "Methods in Chloroplast Molecular Biology" (M. Edelman, R. B. Hallick, and N.-H. Chua, eds.), pp. 259–280. Elsevier, Amsterdam.

Kolodner, R., and Tewari, K. K. (1979). Inverted repeats in chloroplast DNA from higher plants. *Proc. Natl. Acad. Sci. U.S.A.* **76,** 41–45.

Krebbers, E. T., Larrinua, I. M., McIntosh, L., and Bogorad, L. (1982). The maize chloroplast genes for the β and ε subunits of the photosynthetic coupling factor CF_1 are fused. *Nucleic Acids Res.* **10,** 4985–5002.

McIntosh, L., Poulson, C., and Bogorad, L. (1980). Chloroplast gene sequence for the large subunit of ribulose bisphosphate carboxylase of maize. *Nature (London)* **228,** 556–560.

Ohyama, K., Fukuzawa, H., Kohchi, T., Shirai, H., Sano, T., Sano, S., Umesono, K., Shiki, Y., Takeuchi, M., Chang, Z., Aota, S., Inokuchi, H., and Ozeki, H. (1986). Chloroplast gene organization deduced from complete sequence of liverwort *Marchantia polymorpha* chloroplast DNA. *Nature (London)* **322,** 572–574.

Palmer, J. D. (1982). Physical and gene mapping of chloroplast DNA from *Atriplex triangularis* and *Cucumis sativa. Nucleic Acids Res.* **10,** 1593–1605.

Palmer, J. D. (1983). Chloroplast DNA exists in two orientations. *Nature (London)* **301,** 92–93.

Palmer, J. D. (1985). Comparative organization of chloroplast genomes. *Annu. Rev. Genet.* **19,** 325–354.

Palmer, J. D., and Thompson, W. F. (1981). Rearrangements in the chloroplast genome of mung bean and pea. *Proc. Natl. Acad. Sci. U.S.A.* **78,** 5533–5537.

Schleif, R. F., and Wensink, P. C. (1981). "Practical Methods in Molecular Biology." Springer-Verlag, Berlin and New York.

Shinozaki, K., Ohme, M., Tanaka, M., Wakasugi, T., Hayashida, N., Matsubayashi, T., Zaita, N., Chunwongse, J., Obokata, J., Yamaguchi-Shinozaki, K., Ohto, C., Torazawa, K., Meng, B. Y., Sugita, M., Deno, H., Kamogashira, T., Yamada, K., Kusuda, J., Takaiwa, F., Kato, A., Tohdoh, N., Shimada, H., and Sugiura, M. (1986). The complete nucleotide sequence of the tobacco chloroplast genome: Its organization and expression. *EMBO J.* **5,** 2043–2049.

Southern, E. M. (1975). Detection of specific sequences among DNA fragments separated by gel electrophoresis. *J. Mol. Biol.* **98,** 503–517.

EXPERIMENT

5A RNA Isolation from Light- and Dark-Grown Seedlings

Introduction

Isolation of pure, undegraded, high molecular weight RNA from plant tissue can be difficult because of the high levels of ribonucleases, polysaccharides, pectins, and polyphenols often encountered. It has been found, however, that employing a high pH (≧8.5) in the extraction buffer and disrupting the tissue at low temperatures permit extraction of reasonable yields of undegraded plant RNA (e.g., Davies *et al.*, 1972; Cashmore, 1982).

The key step in RNA purification is the removal of proteins. This is most commonly accomplished by extracting cell homogenates containing RNA with phenol and/or chloroform. In such an extraction the nucleic acids partition into the aqueous phase and proteins partition into the more dense phenol/chloroform phase and to the interface between the aqueous and organic phases (Kirby, 1957). A standard phenol extraction to remove proteins from nucleic acid solutions typically involves one extraction with phenol or a 1:1 mixture of phenol and chloroform, and one or more extractions with chloroform. This procedure takes advantage of the fact that deproteinization is more efficient when two different organic solvents are used instead of one. Phenol alone is not the most effective extraction solvent for RNA because, although it readily denatures proteins, it does not completely inhibit RNase activity (ribonucleases are notoriously stable enzymes). In addition, molecules containing poly(A) tracts are somewhat soluble in phenol (Brawerman *et al.*, 1972). Both of these problems can be circumvented by substituting phenol/chloroform (1:1) for phenol in the extraction procedure, especially in experiments designed to purify high molecular weight messenger RNAs. The final extractions with chloroform also aid to remove traces of phenol from the nucleic acid preparation, which can interfere with subsequent manipulations.

There are a number of further modifications that have been introduced in phenol extraction protocols over the years. One variation includes a small amount of isoamyl alcohol in the phenol/chloroform mixture to help reduce the amount of material found at the aqueous/organic interface in the phase separations. A second common modification involves degrading most of the extracted protein with proteolytic enzymes, such as proteinase K or pronase, which are active against a broad spectrum of native proteins, prior to extracting with organic solvents. These modifications help to increase the yield of nucleic acids and maintain their integrity, depending on the tissue or organism from which they are extracted.

In this exercise, total nucleic acids will be extracted from 7-day-old light- and dark-grown barley or pea seedlings, by using a very simple phenol extraction protocol. These plants are particularly well characterized in terms of their photomorphogenic (light-induced development) responses (e.g., Apel and Kloppstech, 1978; Kaufman *et al.*, 1984). These extracts will be used to prepare high molecular weight RNA fractions, which will be examined for functional integrity by *in vitro* translation in a later laboratory period.

WARNING ***Phenol is caustic! Wear protective eyeglasses, laboratory coats, and gloves when phenol is being handled!*** **If phenol comes into contact with your skin,** ***WASH THE AFFECTED AREA IMMEDIATELY WITH SOAP AND WATER.*** **Also note that** ***PHENOL HAS A NOXIOUS ODOR*** **and should be dispensed and worked with in a fume hood as much as possible.**

Protocol

1. Harvest barley seedlings by cutting them just above the soil line with a clean razor blade. If pea seedlings are used, harvest only leaf (light-grown) or bud (dark-grown) tissues. Harvest the dark-grown seedlings in the dark room under green safe lights. Weigh the tissues; you should use about 2 g. Rinse the leaf tissue in a beaker containing distilled water. The dark leaves should be blotted dry and quick frozen by placing them in a Dewar flask containing liquid N_2. They can then be removed from the dark room and processed side by side with the light-grown plants. Collect a similar amount of leaves from light-grown barley seedlings.
2. Place the leaves in a mortar and carefully freeze them by covering them with liquid nitrogen. As the liquid nitrogen boils off, begin grinding the leaves with the pestle. When only a tiny amount of liquid N_2 remains, grind more vigorously until the tissue is finely powdered.

3. Using a sterile spatula, transfer the powder to a blender jar containing 20 ml of RNA extraction buffer and, in the fume hood, add 20 ml of phenol:chloroform:isoamyl alcohol. Cover the jar tightly by wrapping it with about three to four layers of plastic wrap (Saran wrap works well for this step) and forcing the blender jar lid onto the top of the jar to form a tight seal (this will prevent phenol from leaking through the lid of the blender jar). Blend for 2 minutes at top speed.
4. Decant the homogenate into a 50-ml screw-capped centrifuge tube. Cap the tube tightly and further seal the cap using Parafilm, and shake or vortex the tube for 10 minutes. Note that the blender jar can be used many times. Simply rinse it with sterile water after each homogenization.
5. Separate the aqueous and organic phases by centrifugation at 5000 *g* for 20 minutes at 20–25°C (about 6000 rpm in a Beckman JS7.5 rotor). In order to conserve centrifuge time, perform this centrifugation with another lab group.
6. Using a sterile 5-ml pipet equipped with a pipetting device, carefully transfer the upper, aqueous phase to a fresh 50-ml centrifuge tube. Take care to transfer as little of the thick interface as possible. Add 20 ml of chloroform:isoamyl alcohol to the tube, cap it tightly, and shake or vortex the tube for 5 minutes. Discard the used chloroform:isoamyl alcohol in the waste solvent container in the hood.
7. Separate the phases by centrifugation as in step 5, but reduce the time to 5 minutes.
8. Repeat the chloroform:isoamyl alcohol extraction described in steps 6 and 7. Ideally, the organic/aqueous interface following this extraction should be free of flocculent material. If there is still interface material present following the second chloroform extraction, and if time permits, extract the aqueous phase a third time with chloroform:isoamyl alcohol.
9. Precipitate the nucleic acids by adding an equal volume (20 ml) of isopropanol,* capping the tube tightly, mixing the alcohol and water phases by inverting the tube 8–10 times, and incubating the tube in a dry ice/alcohol bath for 20 minutes.

*Student Note □ All nucleic acid precipitations require salt concentrations of at least 0.2 *M*. There is no need to add NaCl in this precipitation because the extraction buffer already contains 0.2 *M* NaCl.

10. Thaw the tube, then collect the precipitated nucleic acids by centrifugation at 5000 *g* for 20 minutes at 4°C.

11. Decant the supernatant liquid carefully into a clean tube. If the pellet was undisturbed, the supernatant may be discarded. Drain any drops of liquid by inverting the tube on a Kimwipe for 3–5 minutes.

12. Reprecipitate the nucleic acids by resuspending the pellet in 5 ml of TE buffer and vortexing it. Add 200 μl of 5 *M* NaCl, followed by 12.5 ml of ice-cold ethanol.

13. Cap the tube tightly, invert several times to mix well, and precipitate the nucleic acids by incubating the tube in ice for 20 minutes.

14. Collect the precipitated nucleic acids as described in steps 10 and 11.

15. Resuspend the pellet in 1 ml of TE buffer. Hold on ice.

16. Withdraw 25 μl of the nucleic acid suspension and transfer it to a 13 × 100 mm glass tube. Add 975 μl of water to this aliquot, mix well, and measure its absorbance at 230, 260, and 280 nm using water as a blank.*

*Student Note □ The A_{260}/A_{280} ratio should be $\geqq 1.8$ for nucleic acid free of protein. The A_{230} should be less than the A_{260} and may be the same as the A_{280}. High A_{230} readings indicate that residual phenol remains in the preparation. This can be removed by an additional ethanol precipitation.

17. Calculate the nucleic acid concentration as follows:

$$(A_{260}) \times (42.5\ \mu\text{g/ml}) \times \text{dilution factor } (1000/25) = \text{nucleic acids } (\mu\text{g/ml})$$

18. Adjust the nucleic acid concentration to 750 μg/ml by dilution with TE buffer. Record the final volume of the nucleic acid solution. Add 0.25 volume of 8 *M* urea, mix, and then add 0.25 volumes of 10 *M* LiCl (for example, if you have 2 ml of solution, add 0.5 ml of urea and 0.625 ml of LiCl), cap the tube, mix well, and incubate overnight in an ice-water bath. This treatment causes all of the high molecular weight RNA to precipitate (as a lithium salt), while most of the tRNA and DNA remain soluble.

19. Collect the precipitated nucleic acids by centrifugation at 5000 *g* and 4°C for 20 minutes. Decant and discard the supernatant, which contains most of the DNA and low molecular weight RNAs. Resuspend the LiCl pellet, which contains the high molecular weight RNA (typically 95–99% rRNA and 1–5% mRNA), in 2 ml of TE buffer (this will probably require a fair amount of vortexing) and remove a 25-μl aliquot for estimating the high molecular weight RNA content. Reprecipitate this RNA by adding 200 μl of 3 *M* potassium acetate (0.1 volume) and 4 ml of cold ethanol (2 volumes). Chill the tube for 20 minutes in a dry ice/ethanol bath. While the RNA is precipitating, estimate the amount of RNA in your 25-μl aliquot by diluting with water and measuring the A_{230}, A_{260}, A_{280} of the diluted sample as described in step 16.

20. Calculate the recovery of RNA as follows:

$$(A_{260}) \times (40\ \mu g/ml) \times \text{dilution factor} \times \text{volume of recovered material (ml)} = \text{RNA recovered } (\mu g)$$

21. Collect the reprecipitated RNA by centrifugation as described in step 19. Decant and discard the supernatant. Invert the tube on several Kimwipes to drain as much of the supernatant as possible (allow about 3–5 minutes). Resuspend the RNA in 2 ml of TE and repeat the ethanol precipitation.*

*Student Note □ This precipitation may be allowed to proceed for "several" days without harming the nucleic acids.

22. Recover the nucleic acids by centrifugation as described in step 19. Invert the tube on a fresh Kimwipe and allow the residual ethanol to drain from the tube for 3 to 5 minutes. Resuspend the RNA in sterile water at a concentration of about 1 mg/ml and store at −70°C.

Materials Provided

All materials and solutions used in the preparation of RNA must be autoclaved. Those which cannot be autoclaved (8 *M* urea) should be made in a sterile graduated cylinder with sterile water.

Barley seedlings	two batches, 5 to 7 days old, one grown in the light and one grown in total darkness
RNA extraction buffer	
50 m*M* Tris–HCl (pH 8.5)	5 ml 1 *M* Tris–HCl (pH 8.5)
10 m*M* EDTA	5 ml 200 m*M* EDTA
200 m*M* NaCl	1.2 g NaCl
	adjust final volume to 100 ml with distilled water
	autoclave

Solution	Recipe
TE 10 m*M* Tris–HCl (pH 7.5) 1 m*M* EDTA	1.21 g Tris base 0.37 g $Na_2EDTA \cdot 2H_2O$ 900 ml H_2O ——— adjust pH to 7.5 with HCl adjust volume to 1 liter with distilled water autoclave
Phenol*	500 g redistilled phenol crystals 200 ml sterile water ——— stir melted phenol crystals and water in a 60°C H_2O bath to make a milky emulsion (~20 minutes); allow to cool to room temperature or 4°C overnight. (Two phases will form: the upper phase is water, the lower phase is water-saturated phenol) store at 4°C in amber or foil-wrapped bottle with bottle cap wrapped tightly with Parafilm

*Student Notes

□ **Caution:** Always wear gloves and work in a fume hood when using phenol. Protective eyeglasses are also recommended. If contact is made with skin, wash affected area with soap and water, *immediately!*

□ To use phenol, dip a pipet through the water layer and withdraw saturated phenol. Always use a pipetting device!

Solution	Recipe
Phenol:chloroform:isoamyl alcohol (50:50:1, v/v/v)†	100 ml phenol, water saturated 100 ml chloroform 2 ml isoamyl alcohol sterile H_2O—enough to cover organic liquid in a dark bottle ——— store at 4°C with bottle cap wrapped with Parafilm

†Student Note

□ **Caution:** The same rules apply to phenol mixtures as outlined above for pure phenol.

Chloroform:isoamyl alcohol (50:1)	mix, saturate with sterile water, and store at room temperature in a dark, tightly capped bottle
8 *M* urea	48 g urea adjust final volume to 100 ml with sterile water (Do not autoclave. Urea breaks down when heated above 55°C)
10 *M* LiCl	42.4 g LiCl adjust final volume to 100 ml with distilled water autoclave
5 *M* NaCl	29.2 g NaCl adjust final volume to 100 ml with distilled water autoclave
3 *M* potassium acetate	29.5 g KCH_3CO_2 (anhydrous) adjust final volume to 100 ml with distilled water autoclave

Mortar and pestle
Capped, 50-ml sterile polypropylene centrifuge tubes
Spatulas/pipets/Pasteur pipets/micropipet tips
Waring blender jar (or Sorvall Omnimixer container)
Dewar flask with liquid N_2

References

Apel, K., and Kloppstech, K. (1978). The plastid membranes of barley (*Hordeum vulgare*). *Eur. J. Biochem.* **85,** 581–588.

Brawerman, G., Mendecki, J., and Lee, S. Y. (1972). A procedure for the isolation of mammalian messenger ribonucleic acid. *Biochemistry* **11,** 637.

Cashmore, A. R. (1982). The isolation of poly A^+ messenger RNA from higher plants. *In* "Methods in Chloroplast Molecular Biology" (M. Edelman, R. B. Hallick, and N.-H. Chua, eds.), pp. 387–392. Elsevier, Amsterdam.

Davies, E., Larkins, B. A., and Knight, R. H. (1972). Polyribosomes from peas. An improved method for their isolation in the absence of ribonuclease inhibitors. *Plant Physiol.* **50,** 581–584.

Kaufman, L. S., Thompson, W. F., and Briggs, W. R. (1984). Different red light requirements for phytochrome-induced accumulation of *cab* RNA and *rbcS* RNA. *Science* **226,** 1447–1449.

Kirby, K. S. (1956). A new method for the isolation of deoxyribonucleic acids: Evidence on the nature of bonds between deoxyribonucleic acid and protein. *Biochem. J.* **64,** 405–408.

EXPERIMENT

5B Preparation of a Wheat Germ Extract for *in Vitro* Translation of mRNA

Introduction

Cell-free extracts prepared from wheat embryos contain ribosomes and soluble factors that will support the translation of mRNAs *in vitro*. The wheat germ translation system is widely used because it is simple to prepare, contains a relatively low level of endogenous mRNA, and will translate most nuclear-encoded eukaryotic mRNAs. In addition, wheat germ is cheap and plentiful. *In vitro* translation extracts from wheat germ have been used successfully to study the synthesis of a wide variety of proteins including those encoded by very abundant or very rare mRNAs and those synthesized in the cytoplasm and transported into chloroplasts. Most current protocols for preparing wheat germ extracts, including the one used in this experiment, are derived from the earlier work of Marcus *et al.* (1974) and Roberts and Paterson (1973). For more details on the preparation of *in vitro* translation extracts, see Erickson and Blobel (1983).

Wheat germ extract preparation involves three basic steps: homogenization of the wheat germ, clarification of the extract by centrifugation, and gel filtration of the clarified supernatant. The last step removes free amino acids from the extract so that radioactively labeled amino acids can be used to follow the course of protein synthesis. All steps should be carried out in a short period of time and the extract should always be kept on ice or in a 4°C centrifuge or cold room. In addition, all buffers used in the protocol are filtered and sterilized to maximize their purity. All glassware, pipet tips, etc., should be autoclaved or heat treated at 150°C to minimize contamination by ribonucleases. Fingers are a rich source of potential nuclease contamination, so gloves should be worn at all times during extract preparation and in assembling the translation reactions.

Special Precautions for Use of Radioisotopes

1. When working with radioactive material, wear lab coats (these should always be left in the laboratory) and disposable vinyl gloves to minimize the chance of personal contamination. These items will be available in the laboratory.
2. Do not pipet by mouth. Most of the operations will employ very small volumes of liquid that must be dispensed with automatic pipetters. If standard pipets must be used, fill them by suction with a pipet-filling device.
3. Work on a square of absorbent paper backed with polyethylene to prevent contamination of the lab bench. Keep all work items on this paper. Mark one corner and designate it as a "hot area." A small radioactive trash container should be kept there as well as all items that are suspected to be contaminated (a small, plastic weighing boat works well for this purpose).
4. Put contaminated, disposable items (e.g., pipet tips, microfuge tubes) into the plastic radioactive trash in your radioactive "hot area." At the end of the period, put this trash and any other contaminated disposable items into the trash can marked "Radioactive Waste." The instructors will indicate the locations of waste containers prior to dispensing any radioactive materials.
5. Put contaminated glassware (e.g., the beaker used for the TCA precipitation assay) in a specially marked dishpan containing Isoclean, a special detergent for decontaminating radioactive labware.
6. After completing work, check your hands and the lab bench for contamination. Use the lab monitor set at the most sensitive setting. It is a good idea to work using the "buddy" system: one lab partner handles the radioactive material; the other lab partner remains "cold" to perform monitoring and handling of nonradioactive solutions. It is also a good idea to wash your hands when you are finished handling radioactive materials, even if no contamination is detected.
7. The organic solvents in liquid scintillation fluid are flammable and toxic when inhaled. Use scintillation fluid in the hood. Cap scintillation vials tightly before leaving the hood area.

Protocols

A. Preparation of Wheat Germ Extract

All steps are performed on ice or in the cold room.

1. Chill two mortars, one in a pan of ice, and the other on the bench top with liquid N_2. **Caution!** Liquid N_2 should be handled carefully to avoid burning the skin. Wear protective gloves, if possible.
2. Add 3 g of wheat germ and some additional liquid N_2 to the second mortar. As the liquid N_2 boils off, begin grinding the wheat germ and continue until it is finely powdered. Add a little more liquid N_2, if necessary, to keep the powder from thawing while you are grinding.
3. Transfer the powder to the mortar on ice. (Use a sterile spatula, and if you lift the liquid N_2-chilled mortar, be careful! It is dangerously cold!) Grind the wheat germ for about 2 minutes with 10 ml of ice-cold homogenization buffer added in three 3.3-ml increments. The material should form a thick paste.
4. Scrape the paste into a 30-ml centrifuge tube.
5. Spin 10 minutes at 23,000 *g* and 4°C (13,000 rpm in the Beckman JA17 rotor).
6. Using a pipet, transfer the supernatant to a fresh 30-ml centrifuge tube. Take care not to transfer any of the pellet and try to minimize carry-over of the floating lipid material.
7. Repeat centrifugation as in step 5.
8. Carefully remove supernatant to a clean tube and crudely measure its volume by aspirating it into a 10-ml pipet.
9. Pass the supernatant through a 1.5 × 30 cm column of Sephadex G-25 in the cold room. Occasionally check the effluent by allowing a few drops to be collected in a beaker containing 10% TCA. Begin collecting effluent in a sterile tube when the first TCA-precipitable material appears. Collect a volume of effluent approximately equal to the volume of supernatant loaded onto the column.
10. Transfer column effluent to a 30-ml Corex tube and centrifuge for 10 minutes at 23,000 *g* and 4°C.
11. Aliquot the translation extract into 500-μl portions in microfuge tubes and quickly freeze in liquid N_2. Store the extract at −70°C.

B. *In Vitro* Translation Reactions

1. Number a series of seven microfuge tubes, put them on ice, and to each one add

    high molecular weight barley RNA, 0 to 6 μl (containing 0 to 6 μg of RNA)
    or
    control mRNA, 6 μl (containing 0.3 μg of mRNA)
    sterile H_2O to bring volume to 6 μl

2. Prepare the energy/amino acid mix (E/AA mix) as follows (in a 500-μl microfuge tube, on ice):

ATP	3 μl
GTP	1 μl
CP	4 μl
CPK	2 μl
−Met AAs	5 μl
[^{35}S]Met	10 μl*
H_2O	25 μl

*Student Note ▫ **Caution:** This is about 100–150 μCi of ^{35}S! Handle with appropriate care.

3. To the E/AA mix, add

compensating buffer	25 μl
wheat germ extract	100 μl

4. To each of the tubes containing RNA and water, add 14 μl of the mixture prepared in step 3.
5. *Gently* vortex the tubes to mix. Spin for 2 seconds in the microfuge to collect all drops in the bottom of the tube.
6. Transfer the tubes to a 25°C water bath and incubate for 90 minutes.
7. Stop reactions by transferring the tube to an ice bath. The tubes can be held here while you prepare materials for the trichloracetic acid (TCA) precipitation assay. After removing duplicate 2-μl aliquots for the TCA assay, store the reactions at −20°C.

C. Trichloroacetic Acid for Measuring Incorporation of Radioactivity into Protein

Precipitation with trichloroacetic acid (TCA) is widely used to separate radioactive biopolymers (proteins or nucleic acids) from radioactive precursor molecules. This separation is accomplished by a simple procedure that precipitates and immobilizes the protein or nucleic acid on a filter. In general, peptides larger than about 5000 molecular weight and oligonucleotides greater than 10 to 20 nucleotides in length are retained on the filter. One of the earliest uses of this assay is described by Mans and Novelli (1961).

1. Prepare a numbered series of 1-cm^2 Whatman 3MM filter papers for the samples to be analyzed. You will need two filters for each sample and two filters for blank controls. The filters should be numbered using a number 2 pencil. Using a P20 Pipetman, pipet duplicate 2-μl aliquots of material to be assayed onto the filters. Be sure to keep a record of which sample is on which filter.
2. Drop the filters into a beaker containing ice-cold 10% trichloroacetic acid (TCA). A large number of filters can be treated in the same beaker. The beaker should be held in an ice-water bath and should contain about 3 to 5 ml of TCA solution for each filter to be analyzed. Hold the filters in the TCA solution for 15 minutes to 1 hour. (This time is not critical and can be as long as overnight, so long as the TCA solution is kept cold.)
3. Decant the first washes of the TCA solution into the designated radioactive waste bottle. Be careful not to decant any of the filters!
4. Wash the filters twice with room-temperature 5% TCA, using about 3 ml/filter. Decant these final washes in the radioactive sink and flush down the drain with copious amounts of water.
5. Heat the filters to a **gentle** boil for 15 minutes in 5% TCA. Perform this step in a fume hood and note that it is **critical** that the filters be boiled gently. **Do not boil off significant amounts of the TCA solution! Boiling the filters in >5% TCA will hydrolyze some of the newly synthesized polypeptides.** Allow the solution to cool for a minute or two in the hood and then discard the TCA in the radioactive sink.

Instructor's Note: Depending on the [^{35}S]Met concentration used in the reactions, the maximum ^{35}S waste generated in this step is about 0.08 × (μCi [^{35}S]Met/μl in original stock) = μCi/filter. Consult local authorities concerning allowable levels of ^{35}S sewer disposal.

Flush the solution down the drain with a large volume of cold water.

6. Wash the filters twice with room-temperature 5% TCA. Discard the washes in the radioactive sink.
7. Wash the filters with 95% ethanol for about 5 minutes at room temperature. Use about 5 ml/filter. Discard the wash fluid.
8. In the fume hood, wash the filters for about 5 minutes in diethyl ether.* Decant the ether into a second beaker in the hood and allow the ether to evaporate. Allow the filters to air dry in the original beaker in the hood for 5 to 10 minutes.

*Student Note ▫ Ether **is extremely flammable** and should be used in a fume hood with the door pulled down for full venting.

9. Place the filters in a numbered set of liquid scintillation vials and add 10 ml of nonaqueous scintillation cocktail. Count the vials in the scintillation counter.†

†Student Note ▫ After counting, the scintillation vial and fluid can be reused a number of times without significantly increasing the background radioactivity by removing the filters with forceps. The radioactive filters are dried in a beaker in a fume hood and disposed of as dry radioactive waste.

10. The net counts per minute (cpm) in each sample are calculated by determining the average cpm for each pair of filters and subtracting the average background cpm on blank filters.

D. Electrophoretic Assay of Newly Synthesized Polypeptides

1. After calculating the average net cpm in each sample, transfer aliquots corresponding to 10^5 cpm to fresh microfuge tubes. As a blank control (for endogenous wheat germ mRNA) use at least half of the zero added RNA control sample.
2. To each sample, add an equal volume of 2× sample loading buffer, incubate in a boiling H_2O bath for 1–2 minutes, then fractionate on an SDS–polyacrylamide gel as described in Experiment 3D.
3. Following electrophoresis, the gel may be stained with Coomassie blue to detect unlabeled marker proteins or di-

rectly processed for fluorography as outlined in Experiment 3D.

4. If 10^5 cpm are available for most samples, fluorographic exposure should be in the range of 1–3 days at −70°C; autoradiography will take about 1 week at −70°C.

Materials Provided

A. Wheat Germ Extract Preparation

2 mortars and pestles
30-ml Corex or polycarbonate centrifuge tubes
Spatulas
Glass or plastic 5-ml or 10-ml tubes

Sephadex G-25 (medium or fine)	swollen and equilibrated in a sterile container with autoclaved water (40 ml/2 groups)
Chromatography columns	1.5 cm × 30–50 cm
Homogenization buffer	
40 m*M* HEPES/KOH (pH 7.6)	10 ml 1 *M* HEPES/KOH (pH 7.6)
100 m*M* potassium acetate	8.3 ml 3 *M* $K(C_2H_3O_2)$
1 m*M* magnesium acetate	0.05 g $Mg(C_2H_3O_2)_2 \cdot 4H_2O$
2 m*M* $CaCl_2$	0.07 g $CaCl_2 \cdot 2H_2O$
4 m*M* dithiothreitol (DTT)	—
	adjust final volume to 250 ml with distilled water autoclave add 0.15 g DTT just before use
Column buffer	
40 m*M* HEPES/KOH (pH 7.6)	40 ml 1 *M* HEPES/KOH (pH 7.6)
100 m*M* potassium acetate	33 ml 3 *M* $K(C_2H_3O_2)$
5 m*M* magnesium acetate	1.1 g $Mg(C_2H_3O_2)_2 \cdot 4H_2O$
4 m*M* dithiothreitol (DTT)	—
	adjust final volume to 1 liter with distilled water autoclave add 0.62 g DTT just before use

10% TCA	10% (w/v) trichloroacetic acid in distilled H_2O

B. *In Vitro* Translation Reactions

ATP	0.1 *M*, neutralized with KOH
GTP	0.02 *M*, neutralized with KOH
Creatine phosphate (CP)	0.6 *M* in 100 m*M* HEPES/KOH (pH 7.6) (Sigma P-6502)
[^{35}S]Methionine	~1000 Ci/mmol, ~10 μCi/μl
Amino acid mix (−Met AAs: all except Met)	1 m*M* each, adjusted to pH 6–8 with KOH
Creatine phosphokinase (CPK)	8 mg/ml, from rabbit muscle (Sigma C-3755); freeze in small (5–10 μl) aliquots. (Use once and discard remainder)
Compensating buffer	
1.0 *M* potassium acetate	16.6 ml 3 *M* $K(C_2H_3O_2)$
5 m*M* magnesium acetate	0.054 g $Mg(C_2H_3O_2) \cdot 4H_2O$
0.8 m*M* spermine	—
20 m*M* dithiothreitol (DTT)	—
	adjust final volume to 50 ml with distilled water autoclave add 0.15 g DTT, 40 μl 1 *M* spermine freeze in 1-ml aliquots

Sterile water
1 mg/ml high molecular weight barley RNA
50 μg/ml control mRNA (e.g., rabbit globin mRNA, BRL #8103SA)

C. Trichloroacetic Acid-Precipitation Assay

Whatman 3MM filter papers, cut into 1-cm squares
Diethyl ether

Solution	Composition / preparation
100% (w/v) trichloroacetic acid (TCA)	500 g TCA 210 ml H_2O add water directly to a fresh bottle of TCA cap the bottle and seal with Parafilm dissolve the crystals by swirling and warming for a few minutes in a 37°C water bath; adjust final volume to 500 ml; store at 4°C
10% (w/v) TCA	100 ml 100% TCA 900 ml H_2O prepare on the day it is to be used; chill to 0 to 4°C before using
5% (w/v) TCA	50 ml 100% TCA 950 ml H_2O prepare on the day it is to be used

References

Erickson, A. H., and Blobel, G. (1983). Cell-free translation of mRNA in a wheat germ system. *In* "Methods in Enzymology" (S. Fleischer and B. Fleischer, eds.), Vol. 96, pp. 38–50. Academic Press, New York.

Mans, R. J., and Novelli, G. D. (1961). Measurement of the incorporation of radioactive amino acids into protein by a filter-paper disk method. *Arch. Biochem. Biophys.* **94,** 48–53.

Marcus, A., Efron, D., and Weeks, D. P. (1974). The wheat embryo cell-free system. *In* "Methods in Enzymology" (K. Moldave and L. Grossman, eds.), Vol. XXX, pp. 749–754. Academic Press, New York.

Roberts, B. E., and Paterson, B. M. (1973). Efficient translation of tobacco mosaic virus RNA and rabbit globin 9S RNA in a cell-free system from commercial wheat germ. *Proc. Natl. Acad. Sci. U.S.A.* **70,** 2330–2334.

EXPERIMENT

6A Screening of Recombinant Phage Libraries with Cloned cDNA Probes

Introduction

Bacteriophage λ has long been a favorite subject for geneticists, and, over the years, considerable information has been accumulated on the organization, function, and regulation of the λ genome. The phage is a double-stranded DNA virus with a genome size of about 50 kb (Figure 6.1). In λ phage particles, the DNA is arranged as a linear duplex with complementary single-stranded ends 12 nucleotides in length (cohesive ends or *cos* sites). After entering an *E. coli* host, the λ DNA circularizes via base pairing of the cohesive ends and is transcribed as a circular molecule (Figure 6.2). In the early phase of infection, one of two alternative replication pathways is chosen. In one pathway, lytic growth occurs, the circular DNA is replicated manyfold, and a number of λ phage gene products are translated. Using the newly made proteins, progeny λ phage are formed, and the host cell lyses, releasing many new infectious phage particles. In a second pathway, lysogenic growth occurs. The phage DNA becomes integrated into the host chromosome and is subsequently replicated and transmitted to progeny bacteria like other *E. coli* chromosomal sequences. The organization of the λ genome and details of its life cycle are reviewed in Brammer (1982) and Maniatis *et al.* (1982).

The usefulness of λ phage as a cloning vector stems from the fact that the central portion (~40%) of its genome is unessential for replication and lytic propagation. This 20- to 25-kb segment of phage DNA can thus be replaced by foreign (eukaryotic) DNA. A number of λ phage vectors, in which the unessential "stuffer" region is bordered by restriction enzyme sites convenient for cloning foreign DNA, have been constructed by standard mutant selection techniques. These restriction sites have also been removed by mutation from the essential λ phage arms

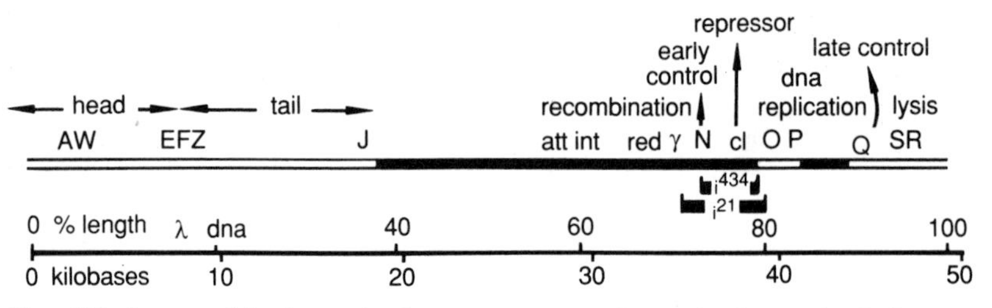

Figure 6.1. Simplified map of the bacteriophage λ genome. Genes in the unshaded regions are essential for phage growth and plaque formation; those in the shaded areas are nonessential. [Details on the expression and function of the different gene clusters are given in Brammer (1982).]

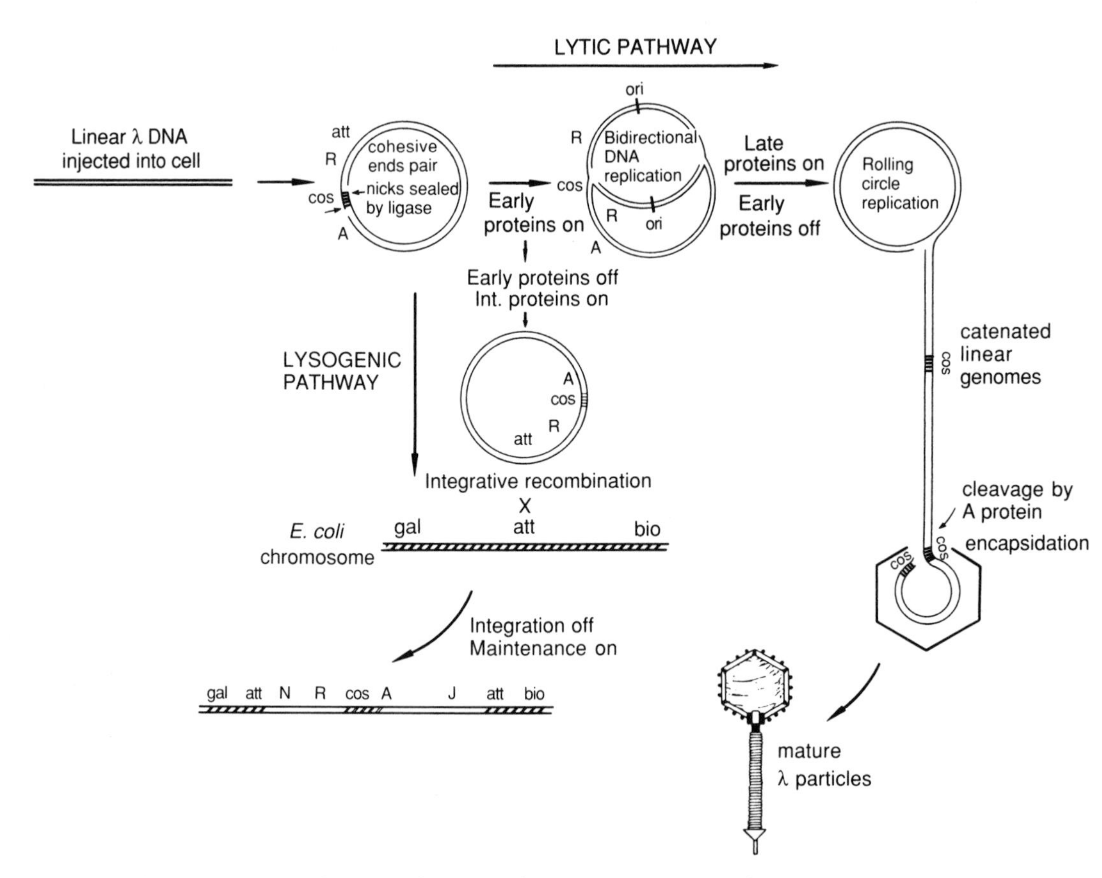

Figure 6.2. A schematic diagram of the bacteriophage life cycle. Double-stranded DNA, which enters the bacterial host as a linear duplex, is indicated by double lines. The lytic pathway, which is indicated by the horizontal arrow, requires early and late gene functions, whereas the lysogenic pathway, which is indicated by the vertical arrow, requires only early gene function. [From Maniatis *et al.* (1982), p. 19.]

which encode head and tail assembly proteins and DNA replication proteins. These manipulations make it simple to substitute the foreign DNA of interest for the λ phage stuffer DNA. A λ phage vector useful for cloning plant genomic DNA is pictured in Figure 6.3.

Libraries of phage containing varying segments of eukaryotic DNA are generally prepared as follows: total cellular DNA is cleaved partially with a restriction endonuclease or by random shearing to a mean size of 17 to 22 kb, the resulting fragments are inserted into an appropriate bacteriophage λ vector, and the recombinant phage are packaged into infectious phage heads *in vitro* (Figure 6.4).

The average angiosperm genome is approximately 4.5×10^9 bp. The exact probability of having any DNA sequence represented in the library can be calculated from the formula

$$N = \frac{\ln(1 - P)}{\ln(1 - f)}$$

where P = the desired probability
f = the fractional proportion of the genome in a single recombinant
N = the necessary number of recombinants

For example, to achieve a 99% probability ($P = 0.99$) of having a given DNA sequence represented in a library of 17-kb fragments, it is necessary to clone and screen the following number of recombinants:

$$N = \frac{\ln(1 - 0.99)}{\ln\left(1 - \frac{1.7 \times 10^4}{4.5 \times 10^9}\right)} = 1.22 \times 10^6 \text{ phage}$$

The technique used to screen such a large number of recombinant phage is plaque hybridization.

In plaque hybridization (Benton and Davis, 1977), a nitrocellulose filter is applied directly to the surface of a plate containing

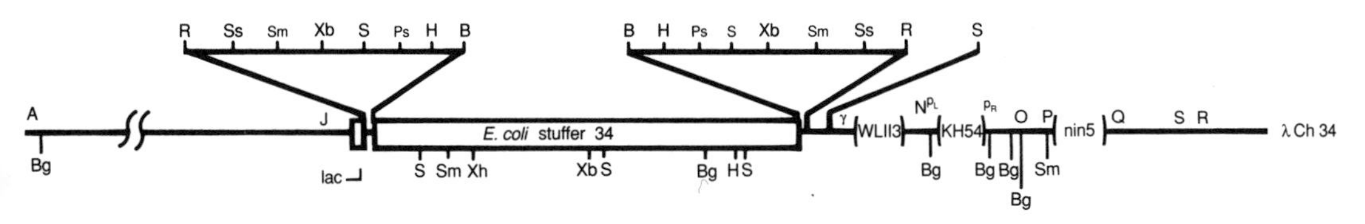

Figure 6.3. Physical and genetic map of phage Charon 34. [Modified from Loenen and Blattner (1983), p. 173.] Restriction sites are abbreviated as follows: B, *Bam*HI; Bg, *Bgl*II; R, *Eco*RI; H, *Hin*dIII; S, *Sal*I; Sm, *Sma*I; Xb, *Xba*I; Ss, *Sst*I; Ps, *Pst*I. The polylinkers were derived from a modified plasmid, pUC 13.

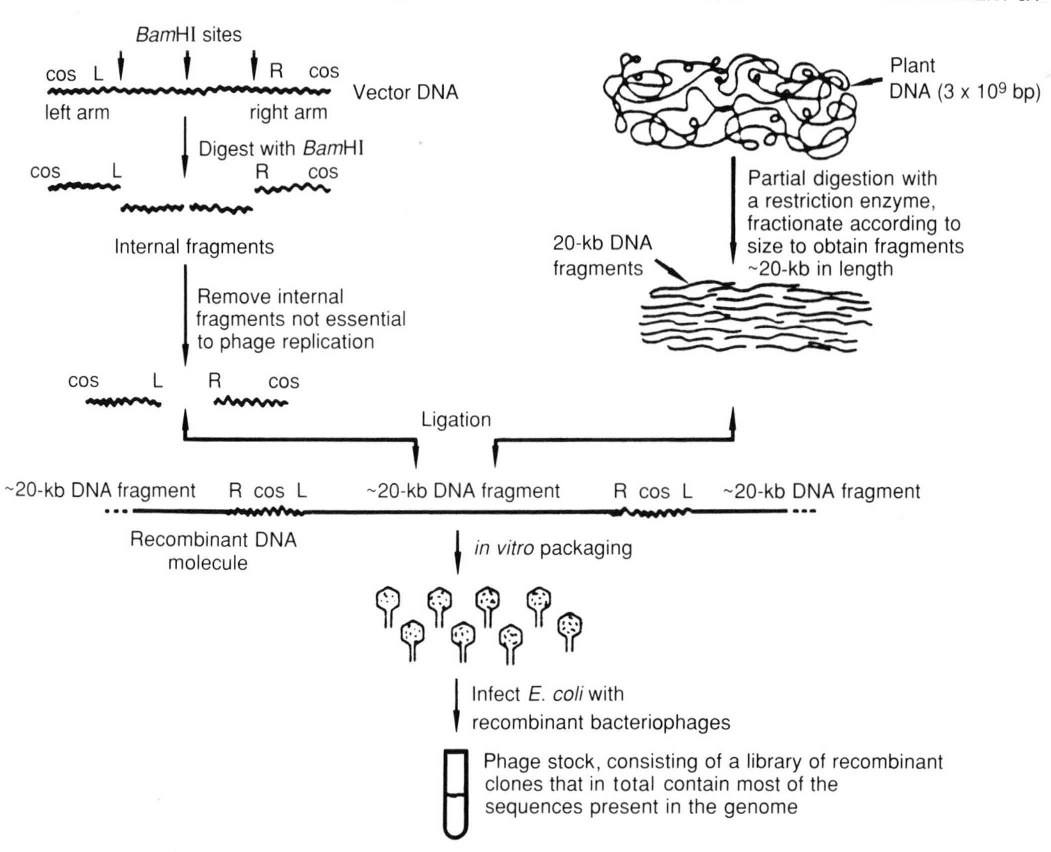

Figure 6.4. A strategy used to construct libraries of random fragments of eukaryotic DNA. *left*, Preparation of the vector DNA fragments; *right*, preparation of eukaryotic DNA fragments. A concatameric recombinant DNA molecule is produced by the action of bacteriophage T4 DNA ligase. This concatamer is the substrate for the *in vitro* packaging reaction during which a different recombinant DNA molecule is inserted into each bacteriophage particle. Following amplification by growth in *E. coli*, a lysate is obtained consisting of a library of recombinant clones that, in aggregate, contain most of the sequences present in the genome. [From Maniatis *et al.* (1982), p. 272.]

bacteriophage plaques. The dry filter is wetted by moisture from the plate. During this wetting, molecules of unpackaged phage DNA present in the lysed plaques bind to the filter. The DNA can be denatured, fixed on the filter, and screened by hybridization with a ^{32}P-labeled DNA or RNA probe (see *Student Note on p. 111) followed by autoradiography just as was done for colony hybridization. Since there is direct contact between the filter and the plaques, the treated filter forms a replica of the pattern of phage plaques and can be used to select a single plaque (containing a single cloned piece of DNA) from an initial collection containing millions of recombinant phage.

In this experiment, we will begin screening a library of plant genomic DNA cloned in λ phage by preparing replica filters of agar plates containing plaques of recombinant phage. In the following periods, we will use these filter replicas to search for sequences encoding rRNA genes which are highly abundant in all eukaryotic genomes (at least 10^3 copies/genome).

*Student Note ▫ If you do not conveniently have a probe for an rRNA gene, this experiment can be done by "spiking" a genomic DNA library with a "positive" phage for which you have a probe.

Protocols

A. Screening of Recombinant Phage Libraries

Day 1

1. Set up five plates of recombinant phage. Each plate is made by mixing approximately 1×10^4 pfu of λ phage with 100 μl of plating bacteria (*E. coli* K802 for Charon 4 libraries; K802 recA$^-$ or DH1 for Charon 34 libraries) in a sterile 13 × 100 mm glass tube. Incubate 10–15 minutes at 37°C.
2. Add 3 ml of T-broth top agarose (cooled to 60°C) to each tube containing bacteria and phage. Rotate the tube back and forth between your palms to mix agarose and phage. Plate the mix immediately on a 100-mm phage plate. Swirl the plate gently so that the mix spreads and covers the entire agar surface. Allow 5–10 minutes for the agarose to harden, then incubate the plate in an inverted position overnight at 37°C.

Day 2

1. Check the plates.† The plaques should fill as much of the plate as possible, but should not be so large as to overlap. Number the plates for future cross reference. Remove plates from incubator and cool at least 30 minutes at 4°C to solidify the top agarose. During this time, set up three trays, one each for denaturation buffer, neutralization buffer, and 2× SSC (~250 ml of each).

†Student Note ▫ Be careful not to turn the plates over with the lids on so that water droplets fall on your phage plaques.

2. Using a ball point pen, carefully number a series of nitrocellulose filters to correspond with the numbering on your plates. Be sure to wear gloves when handling the nitrocellulose. Finger oils will interfere with transfer of DNA to the filter.

3. Remove one plate from the refrigerator and place it right side up on the lab bench with the cover off. Take the correspondingly numbered nitrocellulose filter and, holding the filter on opposite edges with forceps, place the filter directly on the plate. To do this, bend the filter slightly in the middle, and touch this part of the filter to the plate first. Then gradually allow the rest of the filter to touch the surface of the plate so as to trap as little air as possible. **Do not move the filter once it is in place.** Allow about 2 minutes for phage to bind to the filter. During this time, punch three or four holes in an asymmetric pattern through the filter and into the agar using a 20-gauge needle that has been dipped in India ink.
4. **Carefully** peel the filter from the agar surface using blunt-tipped forceps. Wrap the phage-containing plate with Parafilm and store inverted at 4°C.
5. Float the filter (DNA side up) on denaturation buffer for 45–60 seconds. Then immerse the filter for a few seconds.
6. Transfer the filter to neutralization buffer and submerge it for about 5 minutes.
7. Rinse the filter for 5–10 minutes in 2× SSC with occasional gentle shaking.
8. Air dry the filters on paper towels for 20 minutes.
9. Repeat this process with a new filter for each of the plates.
10. Bake filters 2 to 3 hours at 70°C between sheets of 3MM paper.
11. Filters can be stored flat between sheets of 3MM filter paper at room temperature almost indefinitely.

B. Selection of Recombinant Phage by Plaque Hybridization

There are a number of variations that can be employed to hybridize radioactive probes in solution to DNA from phage plaques immobilized on nitrocellulose filters. Details on selecting the correct hybridization conditions have already been presented in Experiment 2.

In this experiment, you will use the replica filters prepared in the previous lab period to set up a hybridization designed to locate recombinant λ phage containing sequences complementary to a ^{32}P-labeled DNA probe derived from a soybean ribosomal RNA (rRNA) gene. If this probe isn't available, you will receive a DNA probe complementary to a positive phage that the library has been "spiked" with.

1. Float the phage replica filters on the surface of a pan of 6× SSC until the filters have wetted (the color will change from white to a grayish color). Submerge the filters and soak in the 6× SSC solution for 5 to 10 minutes.

2. Decant the 6× SSC (directly into the sink, but be careful not to pour out the filters!) and replace it with wash buffer. Incubate the filters, with shaking, for 20 minutes at 42°C.

3a. If using glass crystallizing dishes for the hybridization, transfer the filters one by one to a circular glass crystallizing dish. Allow most of the wash liquid to drain from the filters as they are transferred, but do not let them dry out. Stack the filters one on top of another in the crystallizing dish. Add 60 ml of prehybridization solution, and wrap the top of the dish tightly with Parafilm and cover it with a small glass plate. Incubate 30 minutes with shaking at 42°C.

b. If using seal-a-meal bags for the hybridization, transfer the filters one by one to a seal-a-meal bag, placing them side by side. Add 5 ml of prehybridization solution for each filter (25 ml total) and reseal the bag. Incubate 30 minutes with shaking at 42°C.

4. Prepare hybridization solution as follows: denature the ^{32}P-labeled DNA probe by heating for 5 minutes in a microfuge tube in a boiling water bath (use 1–2 × 10^5 cpm/filter) and add the denatured DNA immediately to the hybridization buffer (60 ml for crystallizing dish; 25 ml for seal-a-meal bag).

5. Decant and discard the prehybridization solution and replace it with the hybridization buffer containing denatured DNA probe from step 4. Note that this step should be done as rapidly as possible in order to minimize drying of the filters. Reseal the crystallizing dish with Parafilm, cover with glass plate, and incubate with gentle shaking overnight at 42°C. (Reseal the seal-a-meal bag and check for leaks before shaking overnight at 42°C.)

6. Remove the hybridization dish or bag from the shaker, remove Parafilm, and decant the hybridization liquid into a suitable radioactive waste container. Immediately replace the solution with about 100 ml of 2× SSC, 0.1% SDS. Cover the dish and swirl gently to separate all the filters. Decant the first wash solution into a radioactive waste container, and transfer the filters to a large tray containing enough 2× SSC, 0.1% SDS to cover them (~300 ml). Wash the filters in this solution with gentle agitation at room temperature for 5–10 minutes. Repeat this wash step two or three more times. The used wash solution that first comes in contact with the filters should be poured into the radioactive waste container; late washes can be poured down the designated radioactive sink in the lab, accompanied by large volumes of water.

7. Wash the filters twice for 30 minutes in 300–500 ml of 0.1× SSC, 0.1% SDS at 50°C. Once again, do not let the filters dry out when changing the wash solutions; discard the washes in the designated radioactive sink.
8. Dry the filters in air on a sheet of Whatman 3MM filter paper at room temperature. You may wish to monitor your filters with a Geiger counter at this point to determine their radioactivity.
9. Tape the filters, numbered side up, to a fresh sheet of 3MM paper. Mark the orientation holes with a small drop of ^{32}P-labeled India ink (your teaching assistant will demonstrate) and also mark the 3MM filter asymmetrically with the labeled ink. These marks will serve as orientation guides that will allow you to align each filter with its respective phage plate.
10. Cover the 3MM sheet and attached filters with plastic wrap. Place the sheet in an X-ray cassette, and in the dark room place a sheet of X-ray film next to the filter side of the 3MM. Close the cassette and expose the film at −70°C. Exposure time will vary, but can range from overnight to about a week.
11. After developing the film, align it with the filters and mark the positions of the orientation holes with a fiber-tipped marker (on the film). Tape a piece of tracing paper to the film and mark the positions of any positive hybridization signals as well as those of the orientation markers. Align the dots on the tracing paper with the marks on the plates to identify the positive plaques.
12. Pick each positive plaque by impaling it with a sterile Pas-

teur pipet and transfer the phage plug to a 13 × 100 mm tube containing 1 ml of SM medium and a drop of chloroform. If a single plaque cannot be selected (which is the usual case in initial screenings) use the wide end of a sterile Pasteur pipet to transfer a large "plug" from the plate to the SM solution. Allow about 30–60 minutes for phage to diffuse out of the plug (at room temperature), then mix aliquots of 1, 5, and 10 μl with 100 μl of plating bacteria. Incubate 10–20 minutes at 37°C, then mix with 3 ml of T-broth top agarose at 60°C, and plate the mixtures on 100-mm phage plates. Grow the phage overnight at 37°C, and, if there is time, repeat the filter replica/hybridization process, using a filter that contains ~100, well-separated plaques. Successive screenings should eventually produce a single pure "positive" plaque.

Materials Provided

A. Phage Screening

K802 cells ($hsdR^-$, $hsdM^+$, gal^-, met^-, supE, rk^-, mk^+)
DH1 cells [F^-, recA1, endA1, gyrA96, thi, hsdR17 (rk^-, mk^+), supE44, re1A1, λ^-]

Denaturation buffer

0.5 *M* NaOH	10.0 g NaOH
1.5 *M* NaCl	43.8 g NaCl
	up to 500 ml with distilled water

Neutralization buffer

0.5 *M* Tris (pH 8.0)	30.3 g Tris base
1.5 *M* NaCl	43.8 g NaCl
	up to 500 ml with distilled water adjust pH to 8.0 with conc. HCl

20× SSC

3 *M* NaCl	350.4 g NaCl
0.3 *M* sodium citrate	178.8 g sodium citrate
	up to 2 liters with distilled water (pH should be 7–7.5)

100-mm phage plates	10 g bacto-tryptone 5 g yeast extract 5 g NaCl 2 g $MgCl_2 \cdot 6H_2O$ 15 g bacto-agar ——— adjust final volume to 1 liter with distilled water autoclave, pour plates store the plates for 1 day at room temperature to dry them out before phage plating. (Do not use plates that are more than 2 days old because they will be too dry for good filter lifts)
T-broth top agarose	10 g bacto-tryptone 5 g NaCl 2 g $MgSO_4 \cdot 7H_2O$ 1 ml 1% vitamin B_1 5 g agarose ——— adjust final volume to 1 liter with distilled water autoclave in 50-ml portions microwave to melt agarose and cool to 50°C just before use

Nitrocellulose filters (82 mm, 0.45-μm pore size)
Flat-tipped forceps
20-gauge needle
India ink
Whatman 3MM filter paper
Small glass or plastic trays

B. Hybridization

6× SSC (6 to 20 dilution of 20× SSC)

Prewashing solution

50 m*M* Tris–HCl (pH 8.0)	6.1 g Tris base
1 *M* NaCl	58.4 g NaCl
1 m*M* EDTA	2 ml 0.5 *M* EDTA (pH 7.5)
0.1% SDS	—

continued on next page

	dissolve components in 800 ml sterile water adjust pH to 8.0 with HCl adjust final volume to 1 liter add 5 ml 20% SDS
Prehybridization solution	
40% formamide	40 ml formamide
5× SSC	25 ml 20× SSC
5× Denhardt's solution	5 ml 100× Denhardt's solution
0.1% SDS	0.5 ml 20% SDS
	adjust final volume to 100 ml with sterile water warm to hybridization temperature
Hybridization solution	
40% formamide	40 ml formamide
5× SSC	25 ml 20× SSC
1× Denhardt's solution	1 ml 100× Denhardt's solution
0.1% SDS	0.5 ml 20% SDS
20 ng/ml ^{32}P-labeled probe*	1 ml 2 μg/ml probe (1–2 × 10^5 cpm/filter)
	adjust final volume to 100 ml with sterile water mix all ingredients, except probe, and warm to hybridization temperature

*Student Note ▫ ^{32}P-Labeled probe is treated just before adding to hybridization solution by incubating in a boiling water bath for 5 minutes (in a sealed microfuge tube) and then adding to warm hybridization solution.

20× SSC	175 g NaCl 88 g sodium citrate
	if necessary, adjust pH to 7.4 with NaOH adjust final volume to 1 liter with distilled water

SM medium

0.1 *M* NaCl	2.92 g NaCl
0.01 *M* $MgSO_4$	5 ml 1 *M* $MgSO_4$
0.05 *M* Tris–HCl (pH 7.5)	25 ml 1 *M* Tris–HCl (pH 7.5)
0.1% gelatin	0.5 g gelatin
	adjust final volume to 500 ml with distilled water
	autoclave in 100-ml portions

100× Denhardt's solution

2% polyvinylpyrrolidone	2 g polyvinylpyrrolidone
2% Ficoll 400	2 g Ficoll 400
2% bovine serum albumin	2 g BSA
	adjust final volume to 100 ml with sterile water
	filter through a 0.2-μm filter
	store at 4°C and never freeze

Plastic or glass trays for washing filters (one or two per lab group)
Blunt-tipped forceps
Glass crystallizing dish (one for the entire class)

References

Brammar, W. J. (1982). Vectors based on bacteriophage lambda. *In* "Genetic Engineering" (R. Williamson, ed.), Vol. 3, p. 53. Academic Press, London.

Benton, W. D., and Davis, R. W. (1977). Screening λgt recombinant clones by hybridization to single plaques *in situ*. *Science* **196,** 180–182.

Clarke, L., and Carbon, J. (1976). A colony bank containing synthetic ColE1 hybrid plasmids representative of the entire *E. coli* genome. *Proc. Natl. Acad. Sci. U.S.A.* **72,** 4361–4365.

Loenen, W. A., and Blattner, F. R. (1983). Lambda Charon vectors (Ch 32, 33, 34, and 35) adapted from DNA cloning in recombination-deficient hosts. *Gene* **26,** 171–179.

Maniatis, T., Fritsch, E. F., and Sambrook, J. (1982). "Molecular Cloning: A Laboratory Manual," p. 270. Cold Spring Harbor Laboratory, Cold Spring Harbor, New York.

Schleif, R. F., and Wensink, P. C. (1981). "Practical Methods in Molecular Biology," p. 145. Springer-Verlag, Berlin and New York.

EXPERIMENT

6B Isolation of Phage DNA from Liquid Cultures

Introduction

In order to give you experience in the isolation of phage DNA after you have obtained a pure stock of a positive phage, you will be provided with a phage stock for DNA isolation and restriction mapping. The most rapid method for isolating and characterizing recombinant phage DNA is a liquid culture method which involves growing the pure phage to very high density in liquid culture, concentrating the phage particles by precipitation with polyethylene glycol, lysing the phage with SDS and heat, and then extracting the DNA with phenol: chloroform. After three ethanol precipitations, this phage DNA can be mapped directly on agarose gels or by Southern hybridization.

Protocol

1. In a 13 × 100 mm tube mix 50 μl undiluted phage suspension with 100 μl fresh overnight culture of K802 or DH1 *E. coli* cells grown in YT broth containing 0.2% maltose. Incubate for 15 minutes at 37°C.
2. Using a sterile Pasteur pipet, inoculate 100 ml of phage broth with this phage suspension. Shake this culture overnight at 37°C. These cultures can be stored at 4°C for a day or two until you are ready for the phage DNA preparation.
3. Add 6 g NaCl per 100 ml phage lysate (final concentration = 1 *M* NaCl). Swirl until salt dissolves. Pellet chromosomal DNA and membranes from lysed cells by centrifuging for 15 minutes at 7700 *g* (7500 rpm in a Beckman JA17 rotor).
4. Pour supernatant into a clean Erlenmeyer, being sure to discard all of the pelleted material. Add 9 g PEG 8000 and swirl occasionally until dissolved. Chill on ice or at 4°C for

1–2 hours. Do not chill much longer than this because the chromosomal DNA will start to precipitate out of solution.

5. Pellet phage particles by centrifuging PEG suspension for 30 minutes at 10,000 *g* (7,500 rpm in a Beckman JS7.5 rotor or 10,000 rpm in a Beckman JA17 rotor). Discard the supernatant and drain tubes on paper towels to get rid of as much polyethylene glycol as possible.
6. Resuspend the phage pellet in 7 ml [50 m*M* Tris–HCl (pH 7.5), 10 m*M* $MgSO_4$].

   Add 1/10 volume 5% SDS (final concentration = 0.5% SDS)
   1/40 volume 0.2 *M* EDTA (pH 7.5) (final concentration = 5 m*M* EDTA)
7. Heat the phage suspension for 15 minutes at 60°C, then let it sit for 15 minutes at room temperature.*

*Student Note ▫ These steps denature the phage particles in the same way that SDS and heat denatured proteins in the SDS–gel electrophoresis protocols.

8. Transfer your sample to a 50-ml capped centrifuge tube and add an equal volume of phenol:chloroform:isoamyl alcohol (50:50:1). Mix sample by lightly vortexing; incubate for 20 minutes at room temperature.
9. Transfer your sample back to a sterile Corex tube and centrifuge for 10 minutes at 13,800 *g* (10,000 rpm in a Beckman JA17 rotor) to separate phases.
10. Collect the aqueous layer; discard the phenol:chloroform layer. Add 1 ml 2 *M* NaCl (final concentration = 0.2 *M* NaCl) and 15 ml ethanol. Incubate on dry ice (−70°C) for 2 hours or in −20°C freezer overnight.
11. Pellet DNA by centrifuging for 15 minutes at 10,000 rpm in a Beckman JA17 rotor. Save pellet.
12. Reprecipitate the DNA in the pellet two more times (three total precipitations) by adding 2 ml 0.2 *M* NaCl and 4 ml ethanol, chilling for at least 15 minutes on dry ice, and recentrifuging for 10 minutes at 10,000 rpm.
13. Dry the DNA pellet and resuspend in 750 μl sterile water.
14. Set up restriction digestions with *Hin*dIII, *Eco*RI, and *Bam*HI, single and double digestions using 10 μl phage DNA per 50 μl restriction digestion. At end of digestions, add 10 μl 10 *M* urea loading dye and load 15 μl of each sample on a 0.8% agarose gel.

Materials Provided

Solution	Preparation
Phage broth	5 g NaCl 5 g yeast extract 8 g bacto-tryptone 2 g $MgCl_2 \cdot 6H_2O$ adjust final volume to 1 liter with distilled water autoclave in 100-ml portions in 250-ml Erlenmeyers
20% maltose	20 g maltose adjust final volume to 100 ml autoclave
Phage extraction buffer 50 m*M* Tris–HCl (pH 7.5) 10 m*M* $MgSO_4$	12.5 ml 1 *M* Tris–HCl (pH 7.5) 0.3 g $MgSO_4$ (anhydrous) adjust final volume to 250 ml with distilled water autoclave
5% SDS	5 g SDS adjust final volume to 100 ml with sterile water
0.2 *M* EDTA	14.9 g $Na_2EDTA \cdot 2H_2O$ add 150 ml water adjust pH to 7.5 with NaOH pellets adjust final volume to 200 ml with distilled water autoclave
Phenol:chloroform:isoamyl alcohol (50:50:1)	50 ml chloroform 50 ml distilled phenol 1 ml isoamyl alcohol saturate this solution with TE buffer and store in a dark bottle at 4°C

2 *M* NaCl	11.7 g NaCl
	adjust final volume to 100 ml with distilled water autoclave in 10-ml aliquots for each group

EXPERIMENT

7 Dideoxy DNA Sequencing

Introduction

One of the most important technologies that has emerged in molecular biology is that of rapid DNA sequencing. There are two methods of DNA sequencing commonly in use: the chemical degradation method of Maxam and Gilbert (1980) and the enzymatic, dideoxy chain termination procedure of Sanger and co-workers (Sanger *et al.*, 1977). Chemical sequencing relies on base-specific, partial degradation of ^{32}P-end-labeled fragments of DNA, coupled with fractionation of the reaction products by high-resolution polyacrylamide gel electrophoresis (Sanger and Coulson, 1978) and autoradiography for detection of a particular DNA sequence. Dideoxy sequencing, on the other hand, takes advantage of the ability of DNA polymerase to faithfully synthesize a complementary radioactive copy of a single-stranded (ss) DNA template using a short ssDNA primer and to randomly incorporate one of the four 2′,3′-dideoxyribonucleoside triphosphates that are analogs of the deoxyribonucleoside triphosphates. The growing DNA chain is terminated once the dideoxy analog is incorporated, because the 3′ end is no longer a substrate for further addition of deoxynucleoside triphosphates. As in the chemical method, the reaction products are fractionated by gel electrophoresis and visualized by autoradiography. The strategy involved in a "dideoxy" sequencing experiment is outlined in Figure 7.1.

Owing to the wide variety of M13 bacteriophage strains that can be used both as cloning vectors and sources of ssDNA and the development of a chemically synthesized, "universal" primer, dideoxy sequencing has come into widespread use; this experiment will employ these methods. The advantages of dideoxy sequencing over chemical degradation are that a high degree of base specificity (accuracy) is obtained, fewer manipu-

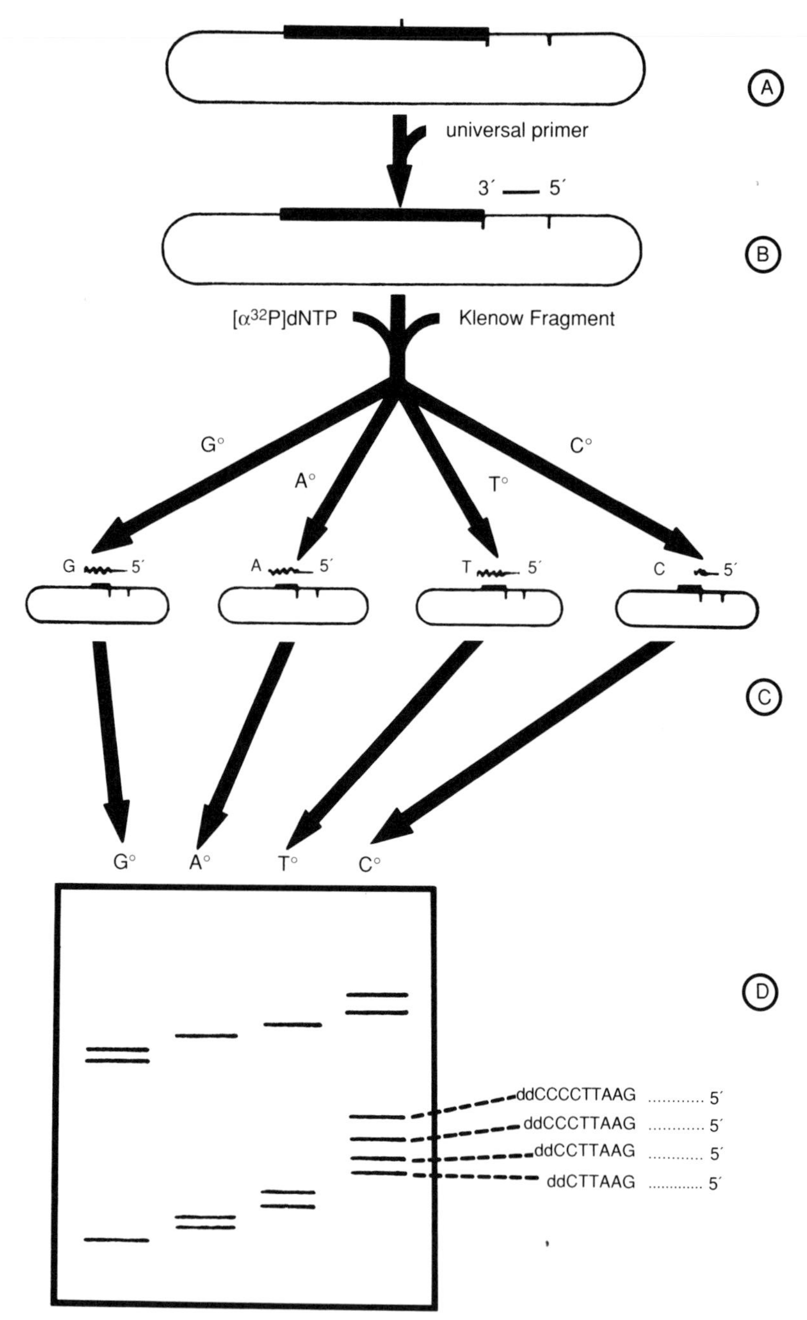

Figure 7.1. Dideoxy DNA sequencing strategy. (A) Single-stranded DNA containing a cloned insert is annealed with a short, primer DNA fragment. (B) The annealed DNAs are mixed with a [^{32}P]dNTP and Klenow fragment and divided into four equal aliquots. (C) Each aliquot is reacted with the remaining three dNTPs and one ddNTP. (D) The four reactions are fractionated by polyacrylamide gel electrophoresis and subjected to autoradiography. (Modified from Bethesda Research Labs, "M13 Cloning/Dideoxy Sequencing Instruction Manual," cover illustration.)

lations are needed, the chemicals used in dideoxy sequencing are far less dangerous than those required for chemical sequencing, and significantly lower amounts of radioactivity are required.

M13 is an *E. coli* single-stranded, filamentous DNA phage (described in detail by Denhardt *et al.*, 1978), with a closed circular genome of about 6500 nucleotides. The life cycle of the phage is outlined in Figure 7.2.

M13 phages attach to F pili of *E. coli* and are therefore able to infect only "male" bacterial cells (F′ or Hfr strains). The ssDNA is converted to the double-stranded replicative form (RF) DNA following penetration into the host cell. This DNA can be isolated from infected cells and used as a cloning vector, much like plasmid DNA. It can be also used to transform *E. coli* by a procedure virtually identical to that used for plasmid DNA (for these procedures, see Messing, 1983). After 100 to 200 copies of RF DNA have accumulated in a cell, DNA synthesis shifts to an asymmetric mode and large amounts of only one of the two DNA strands are produced. These ssDNA molecules are incorporated into mature virions and extruded continually from the host cell. Unlike λ phage infection, M13 infection does not result in lysis and cell death for *E. coli;* host growth is inhibited and the phage produce turbid plaques on lawns of bacterial cells. The

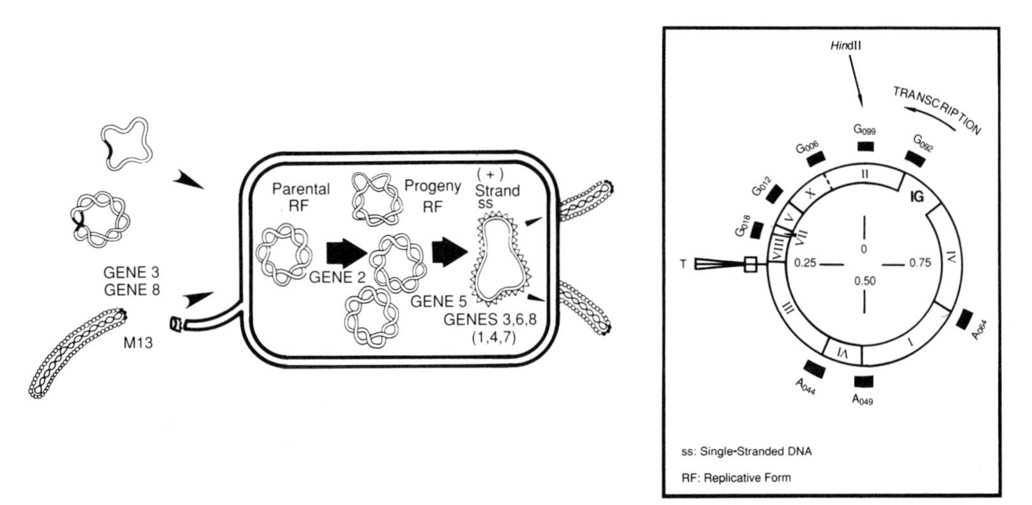

Figure 7.2. Life cycle of M13 phage. Viral genes are designated with roman numerals, and their map positions are indicated on the physical map. The gene products are designated with arabic numerals, and their roles in the various steps of the life cycle are indicated by associating them with the particular intermediate viral form. [Modified from Messing *et al.* (1983), p. 24.]

utility of M13 as a cloning vector for dideoxy sequencing stems from the fact that the phage extruded from the host contain ssDNA that is derived from only one of the two complementary strands of the RF DNA. It is, therefore, a convenient template for dideoxy sequencing reactions.

Nearly all of the M13 genome contains information essential for phage replication, with the exception of a 507-nucleotide region, known as the intergenic sequence (denoted as IG in Figure 7.2). Foreign DNA can be inserted into this region without disrupting the normal phage life cycle. Several M13 strains have been constructed to facilitate cloning and sequencing. The most widely used are the M13 mp series vectors developed by Messing and co-workers (e.g., Yanisch-Perron *et al.*, 1985). These contain two types of DNA insertions in the intergenic sequence:

1. A fragment of the *E. coli lac* operon containing the regulatory region and the coding sequence for the amino-terminal 146 amino acids of β-galactosidase (*lac* Z). This portion of *lac* Z is able to complement a defective *lac* Z gene on an F episome in the host cell whose chromosomal *lac* operon has been deleted ("α complementation"). This complementation permits the transformed host to produce active β-galactosidase. When the phage and cells are grown in the presence of the inducer isopropyl thiogalactoside (IPTG) and the chromogenic substrate 5-bromo-4-chloro-3-indolyl-β-D-galactoside (X-Gal or XG), the active β-galactosidase gives a blue plaque color.
2. A small DNA fragment, referred to as a polylinker, containing several unique restriction sites for cloning. The polylinker is inserted into the amino-terminal portion of *lac* Z. This insertion does not affect α complementation, but additions of foreign DNA into the polylinker region normally destroy the production of β-galactosidase. Phage that harbor foreign DNA in the polylinker region yield colorless plaques when grown in media containing IPTG and X-Gal. The DNA sequence of the polylinker region of a typical M13 cloning vector, one of the mp series constructed by Messing and co-workers (Messing, 1983), is depicted in Figure 7.3.

Generally, M13 vectors differ in the number and/or orientation of the restriction enzyme sites found in the polylinker region. Figure 7.3 also indicates the region of the mp series vectors where small, specific DNA fragments can anneal to prime the DNA sequencing reactions. These primer fragments range from

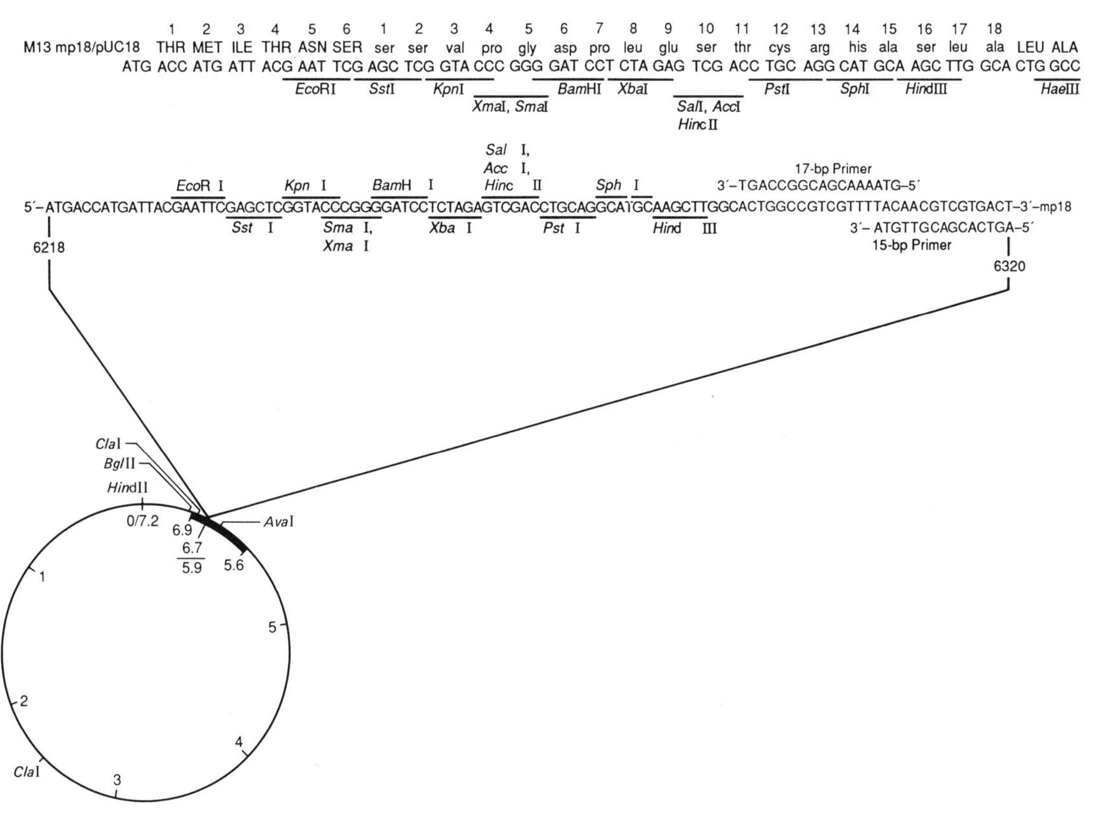

Figure 7.3. An M13 cloning/sequencing vector developed by J. Messing and co-workers (Yanisch-Perron *et al.*, 1985). This vector is one of a pair (pUC 18/19) in which the partner contains an identical "polylinker" in opposite orientation. This arrangement makes it possible to clone an asymmetric DNA fragment in both orientations with respect to the primer annealing site and facilitates sequencing on either DNA strand. [Modified from Maniatis *et al.* (1982), p. 53.]

15 to 26 bases in length. They can be synthesized chemically (Rothstein *et al.*, 1980) and are available commercially.

Both dideoxy and chemical sequencing methods produce "nested sets" of DNA fragments. That is, all the molecules have one end in common (in this case, they are formed from a defined primer) and another end that varies in length. In order to deduce a nucleotide sequence, these fragments must be resolved by gel electrophoresis under conditions in which electrophoretic migration is a function of polynucleotide chain length, and not due to charge differences attributable to base composition. These conditions are achieved at a pH of about 8, where charge differences between the four nucleotide bases are negligible. In addition, the samples are heated in the presence of denaturing agents (formamide in the sample loading buffer and urea in the gel) both before and during electrophoresis. This serves to dissociate the sequencing reaction products from the template DNA and to prevent reassociation between complementary DNA strands and formation of hairpin loops. In both cases, base pairing will impair proper sequence determinations. Following electrophoresis, the acrylamide gel is exposed to X-ray film, which will form an image reflecting the pattern of the fractionated, radioactive DNA molecules.

Deducing a DNA sequence from the autoradiographic pattern simply involves noting the reaction mixture that produces a radioactive band, sequentially, along the length of the gel, starting from the bottom (the shortest reaction products). Because the

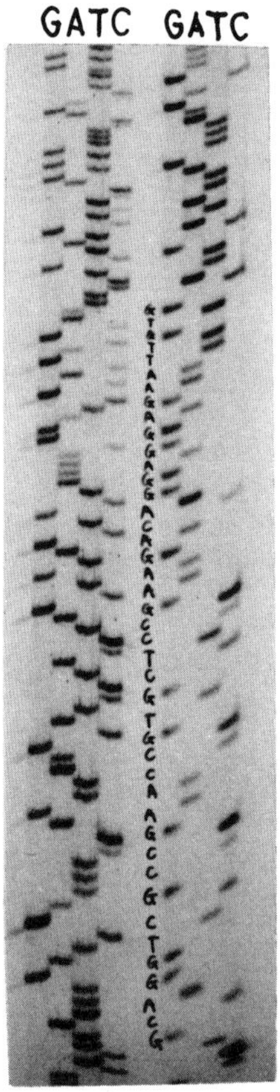

Figure 7.4. Reading a DNA sequencing "ladder." The autoradiograph shows the results of a dideoxy DNA sequencing experiment on a cDNA clone encoding barley leaf calmodulin, a Ca^{2+}-binding protein (V. Ling and R. Zielinski, unpublished). The sequence deduced from the reactions displayed in the right-hand panel is shown in the figure.

reaction products are synthesized in the 5′ → 3′ direction, the nucleotide sequence is read 5′ → 3′ in proceeding from the bottom to the top of the autoradiograph as shown in Figure 7.4.

This experiment presents a basic dideoxy DNA sequencing protocol together with an alternative method that can be used with supercoiled plasmid DNA. Both procedures are based on the formulations for sequencing published by Sanger *et al.* (1977) and Chen and Seeburg (1985). However, it should be noted that improvements are continually being made in these procedures, both to increase sequencing accuracy and to improve the DNA sequence lengths that can be deduced per experiment. These modifications include the use of modified nucleotide bases (dITP) and different DNA polymerases (reverse transcriptase and T7 DNA polymerase). This experiment is meant to acquaint students with the most basic dideoxy sequencing methods, after which they will be prepared to tackle any of the later modifications.

Protocols

A. Preparing M13 Phage Stocks in XL1-Blue

Instructor's (and Student's) Note: This experiment is designed to allow dideoxy sequencing to be set up from scratch. Many of the reagents (particularly the critical nucleotide mixtures) are available commercially in lot-tested kit form. These are available from Bethesda Research Labs, Inc., Promega Biotec, and others. If possible, it is recommended that a sequencing kit be purchased to carry out these procedures, particularly if experience with sequencing is limited or if preparation time is limited.

Day 1

Prepare an overnight culture of XL1-Blue in 2× YT medium containing 12.5–15 μg/ml tetracycline. Grow at 30°C.

Day 2

1. Dilute the overnight culture 1 → 20 with fresh 2× YT containing tetracycline and grow for 1 hour at 37°C.
2. Transfer 2 ml of the XL1-Blue culture into a 50-ml, capped, conical centrifuge tube. Add a plaque of M13 phage by transferring with a toothpick or by picking with a Pasteur pipet.
3. Shake *vigorously* (300 rpm in the New Brunswick shaker) at 37°C for 3 hours.
4. Pellet the cells by centrifugation at 1500 *g* for 15 minutes.
5. Decant the supernatant to a clean tube. Save this as a phage stock at 4°C. The pellet may also be saved at 4°C as a stock for the clone.

B. Determining Insert Size of M13 Clones

(For inserts >200 bp)

1. Take 20 μl of phage stock from step A.5 and combine it with 2 μl of 10× DNA gel sample buffer. Heat 5–10 minutes at 50°C.
2. Load the sample on a 0.8% agarose gel and run the gel for ~5 hours at 100 V. Include a sample of nonrecombinant phage from a blue plaque as a control in an adjacent lane of the gel. Include a *Hin*dIII-cut λ DNA standard on the gel.
3. Stain the gel for 15 minutes in 0.5 μg/ml EtBr solution. Destain with water for 5 minutes.
4. Photograph the gel. Phage having DNA inserts greater than 200 bp will migrate more slowly than the nonrecombinant phage vector DNA (blue plaques).

C. Preparing M13 ssDNA Template for Sequencing

1. Use phage stock supernatant prepared as described in Protocol A. If the stock has been stored at 4°C, centrifuge again to remove any bacteria.
2. Transfer 1.3 ml of the stock to a microfuge tube. Add 200 μl of 25% PEG 8000 in 3 *M* NaCl. Mix well. Allow to stand 30 minutes at 4°C.
3. Spin 15 minutes in a microfuge to collect phage precipitate.
4. Carefully decant (and discard into 10% Clorox) the supernatant. Allow solution to drain from the walls of the tube onto a paper towel.
5. Add 250 μl of TES buffer to the tube. Resuspend the pellet by vortexing.
6. Add 250 μl of water-saturated phenol:chloroform:isoamyl alcohol (50:50:1). Vortex 10–15 seconds. Spin 2 minutes in microfuge.
7. Remove and discard (in the organic waste) the lower, organic layer using a micropipetter. Extract aqueous phase with 250 μl of chloroform. Vortex 5–10 seconds. Spin 10 seconds to separate phases.
8. Transfer upper, aqueous phase to a clean microfuge tube. Add 25 μl of 3 *M* sodium acetate and 750 μl ethanol.
9. Incubate the tube for either 30 minutes on ice or overnight at −20°C. Collect the precipitated DNA by centrifuging for 15–30 minutes in the microfuge.

10. Rinse pellet (do not vortex) with 1 ml of 70 or 80% ethanol (chilled to −20°C). Spin 10 minutes in microfuge.
11. Decant supernatant and invert tube on a tissue or paper toweling to drain off excess ethanol. Dry pellet 5–10 minutes in vacuum desiccator or in a well-vented fume hood.
12. Resuspend pellet in 10 μl of 1× TE buffer (pH 7.5).
13. Quantitate the DNA recovery (if desired) by electrophoresis of 1-μl aliquots in an agarose gel together with standards of known concentration or by diluting 5 μl of DNA to 500 μl with H_2O and reading the A_{260} (1 $A_{260} \simeq 40$ μg/ml). Usually 1–2 μl of this solution is sufficient for DNA sequencing.
14. Store the DNA at −20°C.

D. Dideoxy DNA Sequencing Reactions

1. In a 500-μl microfuge tube, combine the following:

1 μg ssDNA template	1–2 μl
4 ng ssDNA primer (17-bp or 15-bp primer)	2 μl
1 μl 10× DNA polymerase buffer	1 μl
Sterile water	5.5–6.5 μl
	12.5 μl

Cap tightly with Parafilm and place in a water-filled 13 × 100 mm test tube held in a boiling H_2O bath. Incubate 5 minutes, then remove the test tube (containing the microfuge tube) to the bench top for ~45 minutes. During this slow cooling, the primer will anneal to the template DNA. (Pour acrylamide gels at this stage.)

2. While the DNAs in step 1 are annealing, mark four microfuge tubes with G, A, T, and C, respectively. Add and hold on ice until the DNAs are ready (recipes are given in Materials Provided section)

tube G: 2 μl G mix
tube A: 2 μl A mix
tube T: 2 μl T mix
tube C: 2 μl C mix

3. When annealing is complete, collect the liquid droplets by spinning 2 seconds in a microfuge and add the following items to the microfuge tube containing the annealed DNA

   1 μl [α-^{32}P]dATP (10 μCi/μl, 400 Ci/mmol) or 2 μl [α-^{35}S]dATP (10 μCi/μl, 500 Ci/mmol)
   1 μl 0.1 *M* dithiothreitol
   1 μl 1 U/μl Klenow fragment of DNA polymerase I (dilute with 1× DNA polymerase buffer just prior to using)

   Mix tube contents (~15 μl) by pipetting in and out of the micropipet tip.

4. Dispense 3-μl aliquots of annealed DNA solution (step 3) to each of the four reaction tubes (G, A, T, and C, from step 2). Mix, spin 1 second in microfuge, and incubate 15 minutes at 30°C. For [^{35}S]dATP, incubate 20 minutes. Add 1 μl of Chase solution to each of the four tubes, mix, spin, and incubate another 15 minutes at 30°C.

5. Stop the reactions by adding 10 μl of Formamide/dye mix. The [^{32}P]dATP-labeled samples may be stored at −20°C for 2 to 3 days or used immediately for DNA sequencing gels. Reactions using [^{35}S]dATP can be stored frozen for at least 2 weeks prior to gel fractionation.

E. Gel Electrophoresis of Sequencing Reaction Products

1. Gels for fractionating sequencing partial reactions are high-resolution, 6 or 8% acrylamide/8.3 *M* urea gels that utilize 1× TBE buffer (100 m*M* Tris, 100 m*M* H_3BO_3, 2 m*M* EDTA, pH 8.3) and a long (≥40 cm) format.

2. Gel plates, combs, and spacers are cleaned and assembled while the gel solution is mixing. Shorter gel plates should be siliconized and cleaned well with ethanol to remove excess silanizing reagent. Before each use both plates should be cleaned with ethanol and dried thoroughly with Kimwipes.

3. The gel solution contains

   8 ml 10× TBE buffer
   16 ml acrylamide stock (38:2)
   40 g urea
   ~20 ml distilled H_2O

   Stir rapidly to dissolve; heating briefly in a 37°C bath helps. Adjust final volume to 80 ml with distilled water. Cool briefly on ice (but not so long that urea precipitates out). Add 0.56 ml 10% ammonium persulfate and 20 μl TEMED. Mix rapidly and pour into gel mold. Insert the well-forming comb and lie the gel flat on the lab bench. Allow to stand at least 1 hour before using and preferably overnight.

4. Remove gel comb and assemble gel in the vertical apparatus. Fill buffer compartments with 1× TBE and flush tops of wells with 1× TBE buffer. Be sure air bubbles underneath the gel are removed (a long 18-gauge syringe needle bent at the end into a U shape is convenient for this operation). Run gel (without samples) for ~1 hour at 75 W (for a 40 cm long × 30 cm wide × 0.4 mm thick gel). This will cause the gel temperature to rise to 50–60°C—warm, but not uncomfortably hot, to the touch.

WARNING **High voltages (~2000 V) are used in this step. Be careful to avoid contact with reservoir buffers. Be certain wire leads to power supply are in good repair and properly protected.**

5. Just before the gel prerun is over (5–10 minutes), incubate the reaction mixtures in a boiling H_2O bath for 2–3 minutes. Chill ~1 minute on ice, then hold at room temperature.
6. Stop current to gel. Disconnect leads to the power supply. Flush out wells with fresh buffer (from reservoir) and immediately load 2–3 μl of each sample. To do this, use an automatic pipetter with specially made flat tips for loading very thin gels or an automatic pipetter with thin glass capillary tips.

WARNING **Do not mouth pipet radioactive samples!**

7. Immediately reconnect the gel to the power supply and run at 75 W until the bromphenol blue dye reaches the bottom of the gel. (In an 8% acrylamide gel, bromphenol blue migrates with DNA molecules about 20 bases in length and xylene cyanol migrates with DNAs of about 80 bases.) This run should take about 1.5–2 hours.
8. After the gel has run, discard the upper reservoir buffer and then remove the gel sandwich from the apparatus and set it flat on the lab bench. **Carefully** discard the lower reservoir buffer in the radioactive sink or drain the reservoir liquid into a designated waste container—check with the instructor; it contains the unincorporated [^{32}P]dATP from the sequencing reactions! Rinse the apparatus with distilled H_2O, discard the rinse in the radioactive sink, and set the apparatus aside to dry.

Instructor's Note: Each set of four sequencing reactions will contribute ~1–3 μCi of radioactivity to the lower gel reservoir. Depending on local regulations for sewer disposal and class size, this may be too large a burden for direct sewer disposal. Check local regulations.

Lift the plates apart by gently inserting a spatula between the plates and prying (carefully!). Make sure the gel adheres to the unsilanized plate. Remove spacers and set aside. Cover the gel with a sheet of 3MM filter paper or an old piece of X-ray film. Smooth the paper over the gel by rubbing gently with a paper towel. Peel off paper (gel should stick to paper) and cover it with a sheet of plastic wrap. Fold the wrap around the paper or film and tape it to the back (the side away from the gel). Place this in an X-ray cassette. In the dark room, place a sheet of 14 × 17 inch X-ray film next to the gel (plastic wrap side) and close the cassette tightly. Expose the film overnight at −70°C. Note that ^{35}S-sequencing gels must be dried prior to exposing to X-ray film due to the shorter path length of the ^{35}S β-particle. In addition, drying improves the resolution of ^{32}P-sequencing gels by diminishing the scattering effect of the gel matrix.

F. Buffer Gradient Sequencing Gels

A modified sequencing gel that incorporates a buffer gradient can be used to improve resolution in sequencing experiments and permit longer sequences to be deduced in a single electrophoretic run (Biggin *et al.*, 1983). This technique was designed to overcome two problems inherent in nucleic acid gel electrophoresis: longer DNA molecules are separated in a progressively poorer manner, making sequencing patterns uninterpretable, and shorter DNA molecules are separated by spacing wider than is necessary for reading a sequence. The problem of resolving longer DNA molecules is overcome typically by an extended electrophoretic run. By migrating through a longer gel distance, the larger molecules attain a better separation. However, this increased distance means that either very long gels must be employed or that multiple loadings of sequencing reactions must be made in order to resolve both the shorter and longer molecules simultaneously.

A convenient, time-saving alternative to long gels or multiple loadings lies in forming a buffer gradient in the lower (anodal) gel region to selectively decrease the spacing between the shorter DNA molecules. The principle, briefly, is that as the buffer concentration increases, electrical resistance per centimeter in the lower region of the gel decreases, and, consequently, the voltage drop per centimeter (the force that drives nucleic acids through the gel) also decreases. As a result, the smaller DNA molecules in the lower region of the gel migrate more slowly, and the spacing between them remains small. At the same time, the longer DNA molecules can be run a longer distance in the gel, and the spacing between large molecules of n and $n + 1$ nucleotides increases. Thus, gradient gels compress the space taken up by the shorter DNA molecules and simultaneously increase the separation between long DNA molecules.

The following recipe is designed for sequencing gels with dimensions of about 40 × 18 × 0.04 cm, but can be modified proportionally to accommodate any size gel format.

1. Prepare the gel mold as described in Section E.2.
2. Prepare the following solutions:

Component	Solution I	Solution II
Acrylamide stock	7.5 ml	1.5 ml
10× TBE	2.5 ml	2.5 ml
Urea	25 g	5 g
60% sucrose	—	1.5 ml
Bromphenol blue	—	A few crystals
Distilled water	21.5 ml	0.8 ml

3. Measure out 33 ml of Solution I and 6 ml of Solution II. To Solution I, add 75 μl 10% ammonium persulfate and 15 μl TEMED. To Solution II, add 3 μl 10% ammonium persulfate and 1 μl TEMED. Mix thoroughly, but rapidly.
4. Immediately draw up into a 10-ml pipet, using a pipetting device, 5 ml Solution I, followed by 5 ml Solution II. Do this carefully so there is as little mixing as possible. Draw three to five air bubbles into the pipet to mix the two solutions slightly.
5. Introduce the mixture into the gel mold by gently pipetting the mixture down the center of the glass plates. When the pipet is emptied, add the remainder of Solution I to fill the gel mold.
6. Insert the well-forming comb and allow the gel to polymerize in a horizontal position. After polymerization is complete, the gel may be used immediately or saved for up to 48 hours in a sealed plastic bag containing wet paper toweling.
7. Set up the gel mold in the gel rig as usual, but **do not prerun the gel.** The reservoir buffer should be 0.5× TBE (i.e., 10× TBE diluted 20-fold).

8. Prepare samples as usual, apply to the gel slots, and electrophorese at 50 W (constant power) for about 1 hour and 45 minutes. The bromphenol blue should be very near the bottom of the gel.
9. Remove the gel and process for autoradiography as described in Section E.8.

G. Dideoxy Sequencing Using Double-Stranded DNA

DNA sequencing with chain-terminating inhibitors can also be performed using plasmid DNA provided that the DNA is denatured to allow for primer annealing (e.g., Chen and Seeburg, 1985). This protocol utilizes the pUC plasmids, described in Experiment 2, largely because the same 15- or 17-bp primer can be used with these vectors as were used with M13 mp series vectors (Yanisch-Perron *et al.*, 1985). However, any plasmid DNA can be used provided that an appropriate oligonucleotide primer is employed.

1. Resuspend 1–2 μg of CsCl-purified plasmid DNA in 8 μl of H_2O. Add 2 μl of 2 *N* NaOH. Cap, mix, and incubate 5 minutes at room temperature.
2. Add 3 μl 3 *M* sodium acetate (pH 5.0) and 17 μl H_2O to the denatured DNA. Mix, then precipitate the DNA by adding 75 μl ethanol.
3. Hold the DNA for 10 minutes on ice, then collect the DNA by centrifuging 15 minutes in a microfuge.
4. Rinse the DNA pellet with 100 μl of 70% (v/v) ethanol (chilled). Dry pellet in a vacuum desiccator or in a fume hood.
5. Resuspend the denatured DNA in

   6 μl H_2O
   1 μl 10× dsDNA polymerase buffer
   3 μl primer DNA (10 ng/μl)
   ———
   10 μl

6. Anneal the primer to the denatured template DNA at 37°C for 1 hour.
7. While the DNAs in step 6 are annealing, mark four microfuge tubes with G, A, T, and C, respectively. Add and hold

on ice until the DNAs are ready (recipes are given in Materials Provided section)

tube G: 3 μl G mix
tube A: 3 μl A mix
tube T: 3 μl T mix
tube C: 3 μl C mix

8. When annealing is complete, add to the microfuge tube containing annealed DNA

   4 μl [α-^{32}P]dATP (10 μCi/μl, 400 Ci/mmol)
   1 μl 5 U/μl Klenow fragment

   Mix tube contents (~15 μl) by pipetting in and out of the micropipet tip.
9. Dispense 3 μl of label/primer/template mix (step 8) to each of the four reaction tubes (G, A, T, and C, from step 7). Mix, spin 1 second in microfuge, and incubate 15 minutes at 37°C. Add 1 μl of Chase solution to each tube, mix, spin, and incubate another 15 minutes at 37°C.
10. Stop the reactions by adding 10 μl of Formamide/dye mix. The samples may be used immediately or stored at −20°C for 2–3 days prior to gel electrophoresis.
11. Samples are denatured and fractionated by electrophoresis as described in Section E above.

Materials Provided

A. M13 Phage Propagation

XL-1 Blue host cells	available from Stratagene (#200268) genotype: endA1, hsdR17 (r_k^-, m_k), supE44, thi, λ^-, recA1, gyrA96, relA1, Δ(lac), [F′, proAB, lacI^{Q}ZΔM15, Tn-10(tet^R)] these cells should be propagated in tetracycline-containing media to selectively maintain their F episome, which can be eliminated by prolonged growth on rich media

Solution	Preparation
2× YT medium	16 g bacto-tryptone 10 g yeast extract 5 g NaCl ――――― adjust final volume to 1 liter with distilled water adjust pH (if necessary) to 7.2 ± 0.2 with NaOH autoclave for plates, add 15 g bacto-agar per liter before autoclaving for top agarose, add 8 g *agarose* per liter before autoclaving and autoclave in 50-ml portions
12.5 mg/ml tetracycline	125 mg tetracycline–HCl 10 ml 50% (v/v) ethanol ――――― store at −20°C in foil-wrapped container add to liquid medium at a concentration of 12.5–15 μg/ml
20 mg/ml IPTG	0.238 g IPTG (Sigma #I5502) 10 ml sterile water ――――― sterilize by filtration through 0.22-μm nitrocellulose store in 1-ml aliquots at −20°C
50 mg/ml XG	100 mg X-Gal (IBI #02260) 2 ml dimethyl formamide ――――― store in a dark container at −20°C in 1-ml aliquots
10× DNA gel sample buffer	5 ml glycerol 2 ml 0.5 *M* Na_2EDTA (pH 7.5–8) 0.5 ml 20% SDS 10 mg bromphenol blue ――――― adjust final volume to 10 ml with sterile water store at 4°C or room temperature

B. M13 Phage Preparation

Reagent	Preparation
PEG–NaCl	
25% PEG	100 g PEG 8000 (Sigma #P-2139)
3.3 *M* NaCl	264 ml 5 *M* NaCl
	dissolve PEG with rapid stirring
	adjust final volume to 400 ml with sterile water
	store at 4°C or room temperature
TES buffer	2.42 g Tris base
	1.17 g NaCl
	2 ml 0.5 *M* Na_2EDTA (pH 7.5–8)
	add 900 ml water, stir rapidly to dissolve
	adjust pH to 7.5 with conc. HCl
	adjust final volume to 1 liter with distilled water
	autoclave
Phenol*	500 g redistilled phenol crystals
	200 ml sterile water
	stir melted phenol crystals and water in a 60°C H_2O bath to make a milky emulsion (~20 minutes); allow to cool to room temperature or 4°C overnight. (Two phases will form: the upper phase is water, the lower phase is water-saturated phenol)
	store at 4°C in amber or foil-wrapped bottle with bottle cap wrapped tightly with Parafilm

*Student Notes ▫ **Caution:** Always wear gloves and work in a fume hood when using phenol. Protective eyeglasses are also recommended. If contact is made with skin, wash affected area with soap and water, *immediately!*
▫ To use phenol, dip a pipet through the water layer and withdraw saturated phenol. Always use a pipetting device!

Reagent	Preparation
Phenol:chloroform: isoamyl alcohol (50:50:1, v/v/v) (see †Student Note on p. 140)	100 ml phenol, water saturated
	100 ml chloroform
	2 ml isoamyl alcohol
	sterile H_2O—enough to cover organic liquid in a dark bottle
	store at 4°C with bottle cap wrapped with Parafilm

†Student Note ▫ **Caution:** The same rules apply to phenol mixtures as outlined above for pure phenol.

3 *M* sodium acetate	246.1 g $NaC_2H_3O_2$ 600 ml distilled water ——— adjust pH to ~6–7 with "a few drops" of acetic acid adjust final volume to 1 liter filter through 0.22-μm nitrocellulose autoclave
10× TE buffer	12.1 g Tris base 50 ml 0.2 *M* Na_2EDTA (pH 7.5–8) 900 ml distilled water adjust pH to 7.5 with HCl ——— adjust final volume to 1 liter autoclave
10× TBE buffer	121.1 g Tris base 61.8 g H_3BO_3 7.4 g $Na_2EDTA \cdot 2H_2O$ 800 ml distilled water ——— check pH, it should be near 8.3 without adjustment adjust final volume to 1 liter filter through 0.22-μm nitrocellulose autoclave

C. Dideoxy DNA Sequencing

17-bp primer DNA (2 μg/ml in H_2O)	store frozen (−20°C)
10× DNA polymerase buffer	70 m*M* Tris–HCl (pH 7.5) 70 m*M* $MgCl_2$ 500 m*M* NaCl
0.1 *M* dithiothreitol	15.4 mg/ml make up with sterile water ——— store frozen at −20°C for no more than 1 month

Nucleotide stocks	10 m*M* solutions of each dNTP and each ddNTP in sterile water
	all solutions are neutralized with Tris base and should be stored frozen (−20°C); they are stable for months
0.5 m*M* dNTP stocks	prepare by dilution from 10 m*M* stock solutions store at −20°C for no longer than 1 month
G mix	1 μl 10 m*M* dTTP 1 μl 10 m*M* dCTP 1 μl 0.5 m*M* dGTP 2.3 μl 10 m*M* ddGTP 20 μl 10× DNA polymerase buffer 96.7 μl sterile H_2O
A mix	1 μl 10 m*M* dTTP 1 μl 10 m*M* dCTP 1 μl 10 m*M* dGTP 4 μl 10 m*M* ddATP* 20 μl 10× DNA polymerase buffer 133 μl sterile H_2O

*Student Note ▫ For ^{35}S sequencing, use 2 μl 10 m*M* ddATP.

T mix	1 μl 0.5 m*M* dTTP 1 μl 10 m*M* dCTP 1 μl 10 m*M* dGTP 6.1 μl 10 m*M* ddTTP 20 μl 10× DNA polymerase buffer 92.9 μl sterile H_2O
C mix	1 μl 10 m*M* dTTP 1 μl 0.5 m*M* dCTP 1 μl 10 m*M* dGTP 3 μl 10 m*M* ddCTP 20 μl 10× DNA polymerase buffer 96 μl sterile H_2O
Chase mix	1 m*M* in each of the four dNTPs

Formamide/dye mix	9.5 ml deionized formamide
	0.5 ml 0.2 *M* EDTA (pH 8.0)
	10 mg xylene cyanol
	10 mg bromphenol blue
	0.3 ml sterile H_2O
	store at −20°C, make fresh monthly

D. Double-Stranded DNA Sequencing

17-bp primer DNA (10 μg/ml in H_2O)	store frozen at −20°C
10× dsDNA polymerase buffer	100 m*M* Tris–HCl (pH 7.5)
	100 m*M* $MgCl_2$
	500 m*M* NaCl
	50 m*M* dithiothreitol
G mix	1 μl 10 m*M* dTTP
	1 μl 10 m*M* dCTP
	1 μl 0.5 m*M* dGTP
	1.8 μl 10 m*M* ddGTP
	20 μl 10× dsDNA polymerase buffer
	126.2 μl sterile H_2O
A mix	1 μl 10 m*M* dTTP
	1 μl 10 m*M* dCTP
	1 μl 10 m*M* dGTP
	9.1 μl 10 m*M* ddATP
	20 μl 10× dsDNA polymerase buffer
	118.9 μl sterile H_2O
T mix	1 μl 0.5 m*M* dTTP
	1 μl 10 m*M* dCTP
	1 μl 10 m*M* dGTP
	3.5 μl 10 m*M* ddTTP
	20 μl 10× dsDNA polymerase buffer
	124.5 μl sterile H_2O
C mix	1 μl 10 m*M* dTTP
	1 μl 0.5 m*M* dCTP
	1 μl 10 m*M* dGTP
	2 μl 10 m*M* ddCTP
	20 μl 10× dsDNA polymerase buffer
	126 μl sterile H_2O

All other solutions are the same as for dideoxy DNA sequencing.

E. Gel Electrophoresis of Sequencing Reaction Products

10× TBE buffer	Same composition as buffer described in the M13 phage preparation section on p. 140
Acrylamide stock (38:2)	
38% (w/v) acrylamide	38 g acrylamide
2% (w/v) bisacrylamide	2 g bisacrylamide
	adjust final volume to 100 ml with distilled H_2O filter through 0.22-μm nitrocellulose store in an amber bottle at 4°C
10% ammonium persulfate	1 g ammonium persulfate
	adjust final volume to 10 ml with distilled H_2O store at 4°C for no longer than 2 weeks

References

Bethesda Research Labs (BRL). "M13 Cloning/Dideoxy Sequencing Instruction Manual."

Biggin, M. D., Gibson, T. J., and Hong, G. F. (1983). Buffer gradient gels and ^{35}S label as an aid to rapid DNA sequence determination. *Proc. Natl. Acad. Sci. U.S.A.* **80,** 3963–3965.

Chen, E. Y., and Seeburg, P. H. (1985). Supercoil sequencing: A fast and simple method for sequencing plasmid DNA. *DNA* **4,** 165–175.

Denhardt, D. J., Dressler, D., and Ray, D. S., eds. (1978). "The Single-Stranded DNA Phages." Cold Spring Harbor Laboratory, Cold Spring Harbor, New York.

Maniatis, T., Fritsch, E. F., and Sambrook, J. (1982). "Molecular Cloning: A Laboratory Manual," p. 270. Cold Spring Harbor Laboratory, Cold Spring Harbor, New York.

Maxam, A. M., and Gilbert, W. (1980). Sequencing end-labeled DNA with base-specific chemical cleavages. *In* "Methods in Enzymology" (L. Grossman and K. Moldave, eds.), Vol. 65, pp. 499–560. Academic Press, New York.

Messing, J. (1983). New M13 vectors for cloning. *In* "Methods in Enzymology" (R. Wu, L. Grossman, and K. Moldave, eds.), Vol. 101, pp. 20–78. Academic Press, New York.

Orstein, D. L., and Kashdan, M. A. (1985). Sequencing DNA using ^{35}S-labeling: A troubleshooting guide. *BioTechniques* **3,** 476.

Sanger, F., and Coulson, A. R. (1978). The use of thin acrylamide gels for DNA sequencing. *FEBS Lett.* **87,** 107–110.

Sanger, F., Nicklen, S., and Coulson, A. R. (1977). DNA sequencing with chain-terminating inhibitors. *Proc. Natl. Acad. Sci. U.S.A.* **74,** 5463–5467.

Rothstein, R. J., Lau, L. F., Bahl, C. P., Nasrang, S. A., and Wu, R. (1980). Synthetic adaptors for cloning DNA. *In* "Methods in Enzymology" (R. Wu, ed.), Vol. 68, pp. 98–109. Academic Press, New York.

Yanisch-Perron, C., Vierra, J., and Messing, J. (1985). Improved M13 phage cloning vectors and host strains: Nucleotide sequences of the M13mp18 and pUC19 vectors. *Gene* **33,** 103–119.

EXPERIMENT

8 Transformation of Leaf Discs with *Agrobacterium*

Introduction

This experiment is set up to introduce you to some of the techniques involved in transforming plants with exogenous DNA and culturing plant tissue cultures. The method we have chosen to use for infecting plant tissue with *Agrobacterium tumafaciens* is the "leaf disc" method developed by Horsch *et al.* (1985). In this method, surface-sterilized leaf discs are incubated with an *Agrobacterium* strain containing a Ti plasmid. The leaf discs are then transferred to a "feeder plate" containing a layer of tobacco suspension cells which supply the nutrients for the cells in the leaf disc during the regeneration process. After a period of time, the leaf discs are transferred to a "selective plate" on which only transformed calli will grow.

The original method (Horsch *et al.*, 1985) utilized a mutant Ti plasmid in which the hormone-biosynthetic genes in the T-DNA region had been replaced with a chimeric gene for kanamycin resistance. Using this strain of *Agrobacterium,* transformed calli and regenerated plants could be selected for on the basis of their kanamycin resistance. After incubation with the *Agrobacterium* strain for 2 days, the leaf discs were transferred to an agar plate containing kanamycin. On these plates, the transformed leaf discs exhibit shoot regeneration within 2–4 weeks and root regeneration when put into an appropriate rooting media. In the experiment described below, you will inoculate tobacco leaf discs with a wild-type *Agrobacterium* strain containing the hormone-biosynthetic genes normally found in the T-DNA region. Because the auxin and cytokinin biosynthetic genes in the T-DNA overproduce these hormones in transformed calli, transformation of leaf discs with wild-type *Agrobacterium* can be monitored by the hormone-independent growth of the tissue. Because of this, after incubation of the leaf discs on "feeder plates" you will transfer the leaf discs to hormone-free media and select for transformed hormone-independent calli. Because of the hormone overproduction in these calli, shoot regeneration will not occur. At the end of the course, you will isolate DNA from some of the transformed tissues and carry out genomic DNA Southern analysis to determine whether the T-DNA region has indeed been integrated into the nuclear genome.

Most laboratory classes do not have the laminar flow hoods needed to keep the tissue culture samples fungus free for the entire semester. **This experiment will work best** if the teaching assistants and/or the students obtain transformed calli at the beginning of the semester and subculture them every few weeks on MS agar plates until there are 15–20 g/group.

Protocols

A. Transformation of Leaf Discs

1. Surface sterilize two tobacco leaves with 5% Clorox for 15 minutes. Rinse thoroughly with sterile, distilled water.
2. Use an autoclaved 6-mm paper punch to punch out discs from edges of the leaves. (The paper punch should be dipped in ethanol and flame sterilized if more than one group uses it.)
3. Put leaf discs onto PSP agar plates with top surface of leaf down in direct contact with medium. Incubate 48–72 hours at 26–28°C under fluorescent lights until 50% of discs have noticeably expanded.
4. Select six of the expanded leaf discs and put leaf discs into a test tube with 2 ml of a freshly grown *Agrobacterium* culture (grown in YT broth overnight at 28°C).
5. Gently shake leaf discs to make sure that all the edges are covered. Incubate overnight at 28°C.
6. Take leaf discs out of the bacterial culture and blot dry on sterile filter paper. Lay discs on top of nurse culture plate containing tobacco cell "feeder layer" and two filter paper discs (for details, see Materials Provided section).
7. Incubate leaf discs on nurse plate for 2–3 days at 28°C.
8. After 2–3 days, transfer leaf discs to a hormone-free carbenicillin culture plate. Continue incubating discs at 28°C.
9. Leaf discs should be transferred to a fresh hormone-free plate every 2 weeks during the semester.

(As the semester progresses, calli should appear around the edges of the leaf discs. See Horsch *et al.*, 1985.)

B. Propagation of Carrot Suspension Cell Cultures

This protocol works well with both carrot and tobacco suspension cell lines and is aimed at familiarizing you with the sterile techniques involved in subculturing of suspension cultures. All flasks, pipets, and solutions must be sterile. **All transfers must be done in the transfer hood** with a gas flame used to sterilize the open mouths of test tubes and bottles.

Fifty-milliliter cell cultures are grown in 125-ml capped Erlenmeyer flasks. The cultures are shaken at 150 rpm in a New Brunswick shaker at 27°C. The carrot cells should be propagated every 7 days and tobacco cells every 4 days by taking 10 ml of a fully grown suspension culture and adding it to 50 ml of sterile MS solution. Suspension cultures grow best when propagated at 10^4 cells/ml. Subculturing at lower cell densities leads to slow growth and death of the cultured cells.

The MS medium that you need for the suspension cultures is composed of MS I, MS II, MB+, 2,4-D, and iron–EDTA (NaFe EDTA) solutions. The MS medium is aliquoted with 50 ml per 125-ml flask so that the cultures are aerated properly when shaken.

1. Flame top of fully grown suspension culture. Remove 10 ml of carrot cells and transfer into a fresh 50-ml MS culture. Set up two cultures like this. Put in 27°C New Brunswick shaker set at 150 rpm.
2. Check cultures periodically over the next week to make sure that they are not contaminated.
3. Subculture cells again after 7 days (4 days for tobacco). This second subculturing should be done about 4 days before starting the protoplast preparations.

C. Preparation of Protoplasts from Suspension Cells

1. Set up a 50-ml culture of tobacco suspension cells in MS medium 4 days before you want to make protoplasts. Remember that *whenever* cultures are open to air, they must be in a transfer hood.
2. Take sample from culture, stain with Evan's blue vital stain, and examine viability and shape of cells under microscope.
3. Centrifuge the 4 day old, 50-ml suspension culture at 100 *g* for 6 minutes (700 rpm in Beckman JS7.5 rotor). Discard supernatant. Add 2–3 volumes of the enzyme mixture to the cell "pellet." Gently resuspend by swirling. Transfer cells and enzyme solution to a deep tissue culture disk (100 × 25 mm).

4. Wrap plates with Parafilm. Incubate in a dark incubator at 28°C for 3 hours to overnight with gentle agitation. (Enzyme treatment can be done without shaking, but gentle agitation helps speed up the enzyme digestions.)
5. Remove 20 μl of cell suspension. Mix with 2 μl methylene blue staining solution. Examine cells under a microscope. Protoplasts will appear spherical. Undigested cells with intact cell walls are rectangular.
6. Transfer protoplast suspension to 50-ml centrifuge tubes and fill tubes with sterile 10% mannitol or 10% mannitol, 0.1% $CaCl_2$ solution. (The 10% mannitol, 0.1% $CaCl_2$ solution works better if you want to maintain protoplasts for long periods of time.)
7. Centrifuge protoplasts for 5 minutes at 100 *g* in a swinging bucket rotor. Pipet supernatant off the protoplasts. Refill 50-ml tube with 10% mannitol, 0.1% $CaCl_2$ mix gently and recentrifuge. Pour supernatant off; repeat mannitol wash a third time if supernatant looks yellow. Try to drain pellet by pipetting off as much liquid as possible. Repeat these washes two or three times until supernatant is clear.
8. Resuspend protoplasts in 30 ml of 10% mannitol in MS medium.
9. Use the microscope to check shape of cells and to take a cell count using a calibrated counting grid:

   (Number of protoplasts in 16 squares)(5000)
   = number of protoplasts/ml

10. Dilute protoplasts to 1.5×10^5 cells/ml using the 10% mannitol in MS medium.
11. Transfer diluted protoplasts into sterile 125-ml Erlenmeyer flasks and shake at 150 rpm at 28°C for several days.
12. At next lab period, stain sample of cell culture with Evan's blue vital stain to determine viability and whether protoplasts have started regenerating their cell walls.

D. Isolation of DNA from Transformed Leaf Cells and Southern Analysis of T-DNA Sequences

In the genomic DNA isolation described below, you will be working with minute quantities of tissue. Be careful not to lose your DNA! This procedure is easily scaled up to accommodate the isolation of nuclear genomic DNA from intact leaves. Because callus has less DNA per gram of fresh weight of tissue, the amount of callus tissue needed for this preparation is larger than that needed for leaf tissue. The volume of grinding buffer for callus is reduced to produce a higher DNA concentration. An alternate procedure for this small-scale isolation of DNA from tissue culture cells has been described by Mettler (1987).

1. Surface sterilize 5–10 g leaf tissue by soaking in a solution of 5% Clorox for 15 minutes. Rinse well with sterile water and blot dry. Weigh the tissue on a top loading balance. If callus tissue is being used, weigh out 15–20 g and do not surface sterilize.
2. Freeze the tissue in liquid nitrogen in a precooled (−20°C) mortar until it is very brittle, about 2 minutes.
3. Grind the tissue to a fine powder in a mortar and pestle containing a little bit of liquid nitrogen.
4. Transfer the frozen powdered tissue to a 100-ml sterile plastic graduated cylinder. **Immediately** add 10 volumes (for leaf tissue) or 5 volumes (for callus) of grinding buffer to the cylinder. Add diethyldithiocarbamate to final concentration of 0.1 *M* (molecular weight = 171 g/mol). Let sit on ice for 5 minutes. Homogenize the tissue with the Polytron set at its highest setting for 5 seconds. Repeat once. **(Be careful not to touch bottom of polytron to bottom of beaker or cylinder; it will eat through plastic or glass.)**
5. Immediately pour the solution through a sterile funnel containing two layers of Miracloth and into sterile 30-ml Corex tubes or 50-ml polycarbonate tubes set in an ice bucket. Wash extract through with a little grinding buffer, if necessary.
6. Pellet nuclei by centrifuging the filtrate at 350 *g* for 10 minutes (1000 rpm in a Beckman JS7.5 rotor). Discard supernatant. Save pellet.
7. Resuspend nuclear pellet in 5 ml ice-cold lysis buffer by shaking gently.
8. Add 5 ml phenol:chloroform (1:1) and shake *gently* for 20 minutes at room temperature. *Do not vortex* the DNA. Centrifuge at 7000 rpm for 5 minutes in a Beckman JS7.5 rotor to separate phases.
9. Use wide-bored Pasteur pipet to collect aqueous phase (top layer) and transfer it to a sterile 30-ml Corex tube. Avoid the interface material.
10. Measure the aqueous phase and precipitate the DNA by adding

2 volumes TE buffer
1/10 volume 2 *M* NaCl
2 volumes ethanol

Mix gently. Chill at −70°C for 30 minutes.

11. Centrifuge at 10,000 rpm for 10 minutes in a Beckman JS7.5 rotor to pellet DNA. Pour off supernatant into sterile Corex tube and save in case the DNA did not precipitate well.
12. Reprecipitate DNA by adding 100 μl 2 *M* NaCl, 900 μl water, and 2 ml ethanol to DNA pellet. Hold on dry ice for at least 15 minutes. Centrifuge at 10,000 rpm for 10 minutes in a Beckman JS7.5 rotor. Discard supernatant. Cover tube containing DNA pellet with Parafilm and dry in desiccator.
13. Dissolve DNA in 200 μl sterile water and store at 4°C in refrigerator. (Freezing and thawing will shear high molecular weight DNA.)
14. Set up restriction digestions of nuclear genomic DNA with *Bam*HI, *Hin*dIII, and *Eco*RI, just as you did for the chloroplast DNA in Experiment 4. Use 20 μl nuclear DNA from the transformed calli per 50 μl reaction. After the restriction digestions are finished, add 10 μl loading dye and load 25 μl/well.
15. Pour a 0.8% agarose gel using a comb with wider wells so that you can load 25 μl per lane.
16. Run gel at 60 mA/gel. Stain gel with ethidium bromide. Photograph gel.
17. Carry out a genomic DNA Southern transfer as described in Experiment 4.
18. Bake your Southern transfer for 1 hour at 65–70°C.
19. Prehybridize your filter as before by mixing

    7.5 ml 10× SSC, 0.4% sarkosyl, 2× Denhardt's solution
    6.0 ml formamide
    1.5 ml sterile water
    ____
    15.0 ml

    Allow mixture to prehybridize with filter for 10 minutes at room temperature. Squeeze out excess liquid.
20. Add 10^6–10^7 cpm of the nick-translated T-DNA probe, prepared by the teaching assistant, to 1 ml of water. Boil probe for 3 minutes in boiling water bath. Mix hybridization solution as follows:

7.5 ml 10× SSC, 0.4% sarkosyl, 2× Denhardt's solution
6.0 ml formamide
0.5 ml sterile water
1.0 ml probe in water

15.0 ml

Add probe to filter. Seal bag. Check for leaks. Hybridize overnight at 40°C.

21. Wash filter as before with four 20-minute washes in 2× SSC, 0.5% sarkosyl at 40°C. Dry filter on paper towels. Cover with Saran wrap and expose to XAR-5 film for 5 days at −70°C.

Materials Provided

A. Transformation of Leaf Discs

Tobacco plants
5% Clorox solution
Paper punch
Agrobacterium culture grown overnight at 30°C in YT broth

PSP agar plates	3% sucrose 0.1 μg/ml naphthalene acetic acid (auxin) 1 μg/ml benzyladenine (cytokinin) 10 ml/liter B-5 vitamin stock 0.8% bacto-agar MS salts (without 2,4-D) (see suspension cell cultures) adjust the final pH of this solution to 5.7 autoclave pour 25 ml into 100-mm Petri plates
Nurse culture plates	3% sucrose 3% mannitol 0.1 μg/ml naphthalene acetic acid 1 μg/ml benzyladenine 10 ml/liter B-5 vitamin stock 0.8% bacto-agar MS salts (without 2,4-D) (see suspension cell cultures)

continued on next page

	adjust the final pH of this solution to 5.7 autoclave medium pour agar into 100-mm Petri plates layer *Nicotiana tabacum* suspension cells over plate overlay with a tight-fitting Whatman filter paper disc (guard disc) and then a 7.0-cm Whatman filter paper disc (transfer disc)
Hormone-free carbenicillin culture plates	3% sucrose 10 ml/liter B-5 vitamin stock 0.8% bacto-agar MS salts --- adjust pH to 5.7 if needed autoclave medium cool to 60°C add 500 μg/ml carbenicillin pour plates
B-5 vitamin stock	10 mg nicotinic acid 100 mg thiamin–HCl 10 mg pyridoxine–HCl 1 g *myo*-inositol --- adjust final volume to 100 ml with sterile water

B. Cell Suspension Cultures

The ingredients for MS medium are outlined below. For small classes, it is much easier to purchase Murashige Minimal Organics Medium powder from Gibco, adding 9 g/liter bacto-agar and 0.4 mg/ml 2,4-D. **All of the solutions for plant tissue culture should be made with doubly (or preferably triply) distilled water.**

All of the MS I and MS II stock solutions should be filter sterilized, using a 0.45-μm nitrocellulose filter, and stored at −4°C. Once the MS medium is made it should be autoclaved only once because repeated autoclaving produces sucrose derivatives that are detrimental to tissue culture cells.

MS medium (pH 5.7–5.9)	Amount/liter
Sucrose	30 g
MS I	100 ml
MS II	10 ml
NaFe EDTA	1 ml
MB^+	10 ml
2,4-D (100 mg/liter or 0.45 m*M*)	4 ml

MS I solution*	g/liter
NH_4NO_3	16.5
KNO_3	19.0
$CaCl_2 \cdot 2H_2O$	4.4
$MgSO_4 \cdot 7H_2O$	3.7
KH_2PO_4	1.7

*Student Note ▫ Salts must be added one at a time in this order and allowed to dissolve before adding the next one.

MS II medium	g/liter
H_3BO_3	0.62
$MnSO_4 \cdot 4H_2O$	2.23
or	
$MnSO_4 \cdot H_2O$	1.56
$ZnSO_4 \cdot 7H_2O$	0.86
KI	0.083
$Na_2MoO_4 \cdot 2H_2O$	0.025
$CuSO_4 \cdot 5H_2O$	0.0025
or	
$CuSO_4$	0.0016
$CoCl_2 \cdot 6H_2O$	0.0025

MB^+ solution	g/liter
Glycine	0.2
Thiamin–HCl	0.01
Nicotinic acid	0.05
Pyridoxine–HCl	0.05
myo-Inositol	10.0

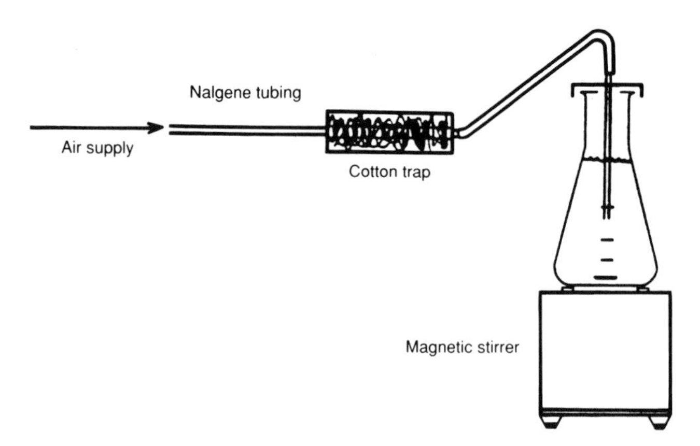

Figure 8.1. Apparatus for NaFe EDTA.

NaFe EDTA solution	g/liter
$Na_2EDTA \cdot 2H_2O$	37.2
$FeSO_4 \cdot 7H_2O$	27.8

Bubble air rapidly through solution for 4–5 hours (until dissolved) at room temperature. Stir constantly. See Figure 8.1.

C. Preparation of Protoplasts

Enzyme solution

5.00 g 1% cellulase*
2.50 g 0.5% hemicellulase†
18.22 g 0.2 *M* mannitol
3.67 g 50 m*M* $CaCl_2 \cdot 2H_2O$
0.68 g 10 m*M* sodium acetate

adjust final volume to 500 ml
adjust pH to 5.8 with 1 *M* NaOH (7 drops/100 ml)
spin 10 minutes at 10,000 rpm
filter sterilize supernatant through a 0.45-μm pore size Millipore filter (Nalga brand #245-0045)

*Student Note ▫ Cellulase "onozuka R-10" is obtained from Yakult Pharmaceutical Industries, Ltd., B-21 Shigikjan-cho, Nishinomiya, Japan.

†Student Note ▫ Hemicellulase (16870) is obtained from United States Biochemical Corporation, Cleveland, Ohio.

10% mannitol	10 g mannitol in total volume of 100 ml
	autoclave
10% mannitol in MS medium	10 g mannitol in total of 100 ml MS medium that has not been previously autoclaved
	autoclave once store at room temperature

2.5% Evan's blue vital stain in 10% sterile mannitol

Protoplast wash buffer	
10% mannitol	50 g mannitol
0.1% $CaCl_2$	0.5 g $CaCl_2 \cdot 2H_2O$
	add 450 ml sterile water adjust pH to 5.7 with 1 *M* NaOH adjust final volume to 500 ml autoclave

D. Isolation of DNA from Transformed Leaf Cells

5% Clorox solution
Liquid nitrogen
Sterile plastic 100-ml graduated cylinders
Sterile mortars and pestles
Miracloth
Polytron
Sterile 30-ml Corex tubes
Sterile 50-ml centrifuge tubes
Diethyldithiocarbamate
TE buffer
2 *M* NaCl
10× *Bam*HI buffer
10× *Eco*RI buffer
10× *Pst*I buffer
10 *M* urea loading dye
Phenol:chloroform (1:1)

Grinding buffer	
0.3 *M* sucrose	153 g sucrose
50 m*M* Tris–HCl (pH 8.0)	75 ml 1 *M* Tris–HCl (pH 8.0)
5 m*M* $MgCl_2$	1.53 g $MgCl_2 \cdot 6H_2O$

continued on next page

	adjust final volume to 1.5 liters with distilled water autoclave
Lysis buffer	
20 m*M* EDTA	10 ml 200 m*M* EDTA
50 m*M* Tris–HCl (pH 8.0)	5 ml 1 *M* Tris–HCl (pH 8.0)
1% sarkosyl	—
	adjust final volume to 100 ml with sterile water autoclave add 1 g sarkosyl per 100 ml

References

Hooykaas, P. J. J., and Schilperoort, R. A. (1983). The molecular genetics of crown gall tumorigenesis. *Adv. Genet.* **22,** 210–283.

Horsch, R. B., Fry, J. E., Hoffman, N. L., Eichholtz, D., Rogers, S. G., and Fraley, R. T. (1985). A simple and general method for transferring genes into plants. *Science* **227,** 1229–1231.

Fraley, R. T., *et al.* (1983). Expression of bacterial genes in plant cells. *Proc. Natl. Acad. Sci. U.S.A.* **80,** 4803–4807.

Mettler, I. J. (1987). A simple and rapid method for minipreparation of DNA from tissue cultured plant cells. *Plant Mol. Biol. Rep.* **5,** 346–349.

Riven, C. J., Zimmer, E. A., and Walbot, V. (1982). *In* "Maize for Biological Research" (W. F. Sheridan, ed.), p. 161. University Press, Grand Forks, North Dakota.

Schell, J. S. (1987). Transgenic plants as tools to study the molecular organization of plant genes. *Science* **237,** 1176–1182.

APPENDIX

I Glossary

Molecular biology, as any rapidly developing technology, has evolved its own peculiar jargon. The following is an informal glossary of some frequently used terms.*

Terms and Definitions

Amplify: (v) To produce multiple copies of a gene, plasmid, or any collection of these molecules.

Blunt: (adj) The ends of linear, double-stranded DNA in which neither the 5′ nor the 3′ strand is protruding, such as, the structure produced by cleavage of double-stranded DNA with *Rsa*I or *Sma*I restriction endonucleases.

cDNA: (n) Complementary DNA produced by primed reverse transcription of an RNA template. cDNA can be single stranded (frequently used as a hybridization probe), or double stranded (used as a method to clone copies of all the mRNA molecules present in a cell at a particular time).

Cleave: (v) Usually used to describe the double-stranded cutting of DNA at specific sites by restriction endonucleases. Syn. **cut, restrict**.

Clone: 1. (v) To isolate and amplify a population of genetically identical organisms or genes. 2. (n) One member of a population of genetically identical organisms or genes.

Coding strand: (n) The mRNA-like strand of double-stranded DNA. It contains the sequence of codons that form the reading frame.

Cohesive end: (n) The end of a linear, double-stranded DNA molecule having a protruding 3′- or 5′-end (or ends). The protruding end(s) can hybridize by base pairing to complementary protruding end(s) on another DNA molecule. Syn. **sticky end**, **overhang**.

Comb: (n) A plastic device used to form sample wells in agarose or acrylamide gels. In vertical gels, used in combination with **spacers** of identical thickness.

3′-End: (n) The end of a single-stranded fragment of DNA or RNA whose deoxyribose or ribose sugar moieties, respectively, possess free 3′-hydroxyl or monophosphate residues.

5′-End: (n) The end of a single-stranded fragment of DNA or RNA whose deoxyribose or ribose sugar moieties, respectively, possess free 5′-hydroxyl or mono-, di-, or triphosphate residues.

Gene: (n) A segment of DNA or RNA that normally contains a promoter, a transcribed region, and a termination signal.

Hybridize: (v) To promote the formation of complementary base pairing between nucleic acid molecules. These may be between two different molecules, or between two different

* v, Verb; adj, adjective; n, noun; Syn, synonym.

regions of the same molecule (e.g., a hairpin loop).

Lawn: (n) A confluent layer of bacteria (usually *E. coli*) grown on an agar plate, which is used to isolate bacteriophage plaques.

Ligate: (v) To join two nucleic acid molecules covalently. Performed *in vitro* by the activity of DNA or RNA ligase.

Linker: (n) A short, chemically synthesized, double-stranded DNA molecule that contains the recognition sequence for a restriction endonuclease. They are ligated to larger, blunt-ended DNA molecules, and then cleaved with the restriction enzyme whose recognition sequence the linker bears. In so doing, cohesive ends are formed on the originally blunt-ended DNA molecule. This process is frequently used to facilitate cloning of blunt-ended DNAs.

Map: 1. (n) A graphic representation of the relative positions of restriction sites or genes on a nucleic acid molecule. 2. (v) To determine the relative positions of restriction sites or genes on a DNA molecule.

Noncoding strand: (n) The strand of DNA that is complementary to the sequence of mRNA in a double-stranded DNA molecule.

Overhang: (n) A single-stranded DNA sequence that protrudes from the 3′- or 5′-end of double-stranded DNA. Syn. **sticky end, cohesive end**.

Plasmid: (n) A circular, double-stranded DNA molecule containing an origin of replication, and selectable marker(s) that frequently confer resistance to various antibiotics.

Plate: 1. (n) A covered glass or plastic dish used to grow bacteria on agar medium or to grow cells in culture. 2. (v) To spread bacteria or phage/bacteria mixtures on an agar medium, often in combination with molten top agar.

Restrict: (v) To subject DNA to the activity of a restriction enzyme. See **cleave**.

Restriction enzyme: (n) A nucleolytic enzyme that recognizes a short, specific DNA sequence (a **restriction site**—usually 4 to 8 base pairs) and cleaves both the DNA strands within or near the same sequence. More formerly termed a **restriction endonuclease**.

Restriction site: (n) The DNA recognition sequence of a restriction enzyme, normally 4 to 8 base pairs in length.

Sense strand: (n) The mRNA-like strand of double-stranded DNA (with the substitution of T for U). The strand that bears the codons encoding the polypeptide produced by that particular gene. Syn. **coding strand, plus (+) strand**.

Spacer: (n) Thin plastic strips used to separate uniformly the glass plates used to cast an acrylamide gel. See **comb**.

Sticky end: See **cohesive end**.

Transform: (v) To alter the genotype of a cell (usually *E. coli*) by adding gene(s) to the cell. Typically, transformation also changes the phenotype of a cell.

Abbreviations

ATP	adenosine triphosphate
bp	base pair
BPB	bromphenol blue
BSA	bovine serine albumin
CP	creatine phosphate
CPK	creatine phosphokinase
2,4-D	(2,4-dichlorophenoxy)acetic acid
dATP	deoxyadenosine triphosphate
dCTP	deoxycytidine triphosphate
ddATP	dideoxyadenosine triphosphate
ddCTP	dideoxycytidine triphosphate
ddGTP	dideoxyguanosine triphosphate
ddNTP	all four dideoxyribonucleotide triphosphates
ddTTP	dideoxythymidine triphosphate
dGTP	deoxyguanosine triphosphate
DNA	deoxyribonucleic acid
DNase	deoxyribonuclease
dNTP	all four deoxyribonucleotide triphosphates
DTE	dithioerythreitol
DTT	dithiothreitol

dTTP	deoxythymidine triphosphate
EDTA	ethylenediaminetetraacetate
EtBr	ethidium bromide
FCCP	*p*-fluoromethoxycarbonylcyanide phenylhydrazone
IPTG	isopropylthiogalactopyranoside
kb	kilobase
Met	methionine
mRNA	messenger RNA
PEG	polyethylene glycol
RNA	ribonucleic acid
rRNA	ribosomal ribonucleic acid
RuBisCo	ribulose-bisphosphate carboxylase
SDS	sodium dodecyl sulfate
SSC	standard saline citrate
ssDNA	single-stranded DNA
TBE	Tris–borate–EDTA buffer
TCA	trichloroacetic acid
TE	Tris–EDTA buffer
TEMED	*N,N,N′,N′*-tetramethylethylenediamine
TES	*N*-Tris(hydroxymethyl)methyl-3-aminopropanesulfonic acid
tRNA	transfer RNA

APPENDIX

II Gel Electrophoresis Equipment

The experiments outlined in this manual are typical of those carried out in most molecular biology laboratories in that they utilize gel electrophoretic fractionation of nucleic acids and proteins extensively. Equipment designed to perform such fractionations can be purchased commercially, often at considerable expense. However, equally serviceable equipment often can be constructed at a fraction of the cost of commercial models, provided that one has access to a reasonably well-equipped machine shop. In this section, we present suggested commercial sources of reliable, moderately priced electrophoresis equipment, as well as plans for constructing both horizontal and vertical gel electrophoresis rigs.

Power Supplies

For horizontal agarose gels and minivertical gels, the basic requirements can be met with a unit that delivers 0–100 mA and 0–200 V. Some very reliable units in this range are

EC model 103 (about $250): 250 V, 100 mA maximum output
EC model 452 (about $750): 500 V, 500 mA maximum output

Both are available from VWR Scientific and Thomas Scientific, often at a considerable discount to educational institutions.

Bio-Rad model 100/200 (about $325): 100 V, 200 mA maximum output

This power supply can be obtained from Bio-Rad Laboratories.

DNA sequencing, unfortunately, requires considerably more voltage capability for optimal results. Typically, around 2000 V are required to create the Joule heating necessary for fully denaturing and resolving DNA fragments differing in length by a single nucleotide. Two reasonably priced and very reliable units that will power up to four sequencing gels simultaneously are the Bio-Rad model 2000/200 (2000 V, 200 mA; about $900) and EC model 600 (4000 V, 200 mA; about $1800).

Electrophoresis Rigs

For laboratory class instruction, commercially built electrophoresis units are normally prohibitively expensive. However, reasonably inexpensive, rugged units are available from Idea Scientific Co., Box 2078, Corvallis, Oregon 97330 [(503) 758-0999]. These include a minivertical slab gel apparatus (about $170), and a horizontal electrophoresis unit (about $160). In this section, we have enclosed plans for constructing horizontal and vertical gel electrophoresis rigs. In each case, the major expense item is the platinum wire used to form the electrodes.

Horizontal Electrophoresis

The plans outlined in Figure AII.1 are for a horizontal electrophoresis apparatus suitable for agarose gel fractionation of DNA restriction fragments. The box is constructed of $\frac{1}{4}$-inch Plexiglas, the well-forming combs are $\frac{1}{16}$-inch polycarbonate, and the electrodes are formed from 26-gauge platinum wire. Note that two combs can be inserted into the box while forming a gel, and thus the sample capacity of the box can be doubled. The ends of the gel-forming box are sealed tightly with adhesive or electrical tape when casting a gel.

Vertical Electrophoresis (Proteins)

The apparatus shown in Figure AII.2 is constructed of $\frac{1}{8}$-inch Plexiglas. Dimensions are shown in millimeters. Glass plates for forming gels are 8 × 10 cm standard lantern slides. They are available from A. H. Thomas Scientific Co., catalog #6686-M20 or 6686-M23. Spacers and combs are made from $\frac{1}{32}$-inch sheets of Teflon by cutting with a single-edged razor blade or an Exacto knife. Spacers should be about $3\frac{1}{2} \times \frac{1}{8}$ inches, and a convenient size for the teeth of the well-forming combs is $\frac{1}{4}$ inch, separated by spaces of $\frac{1}{8}$ inch (10 teeth separated by 9 spaces = $3\frac{5}{8}$ inches).

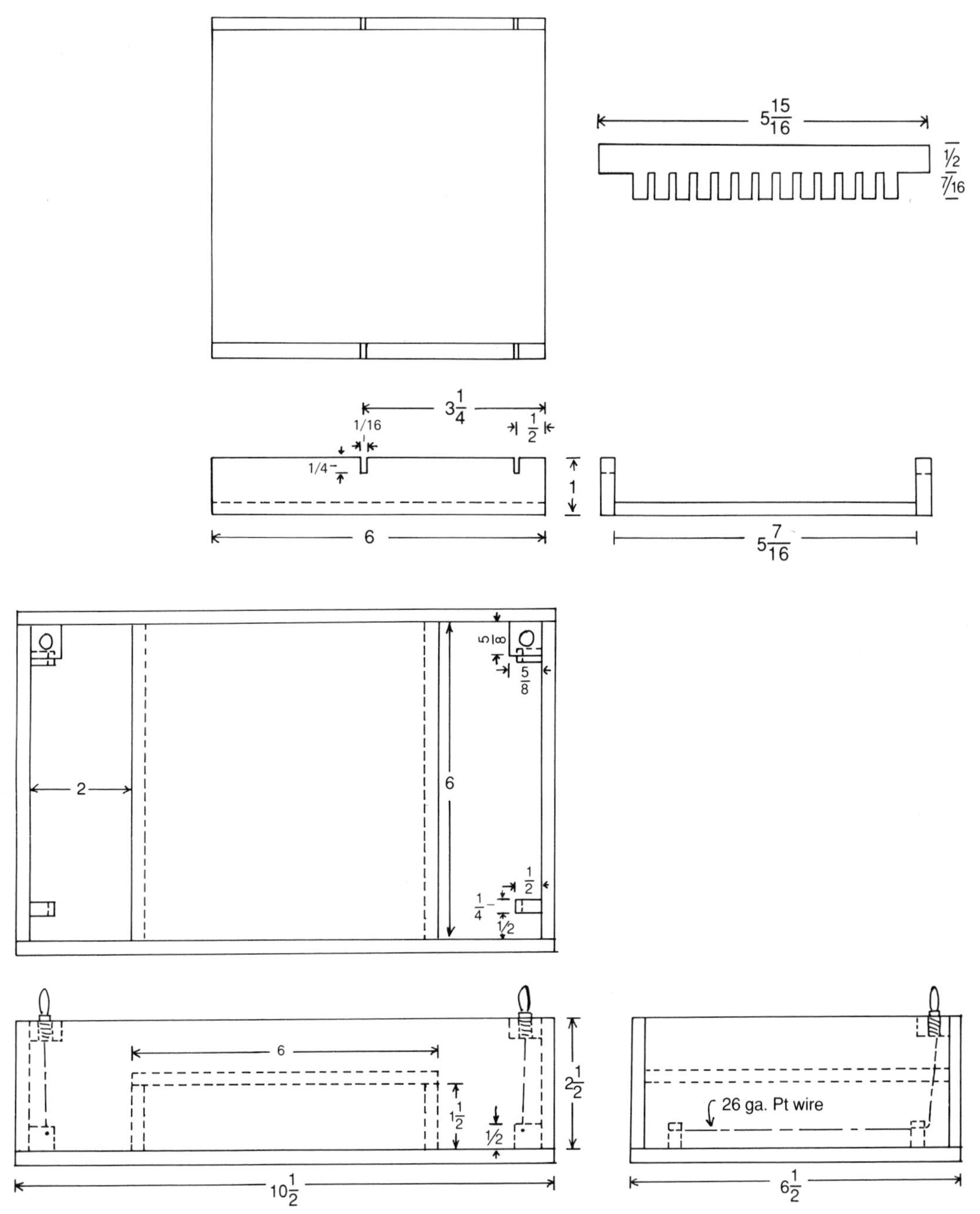

Figure AII.1. Horizontal electrophoresis apparatus suitable for agarose gel fractionation of DNA restriction fragments. All dimensions are in inches.

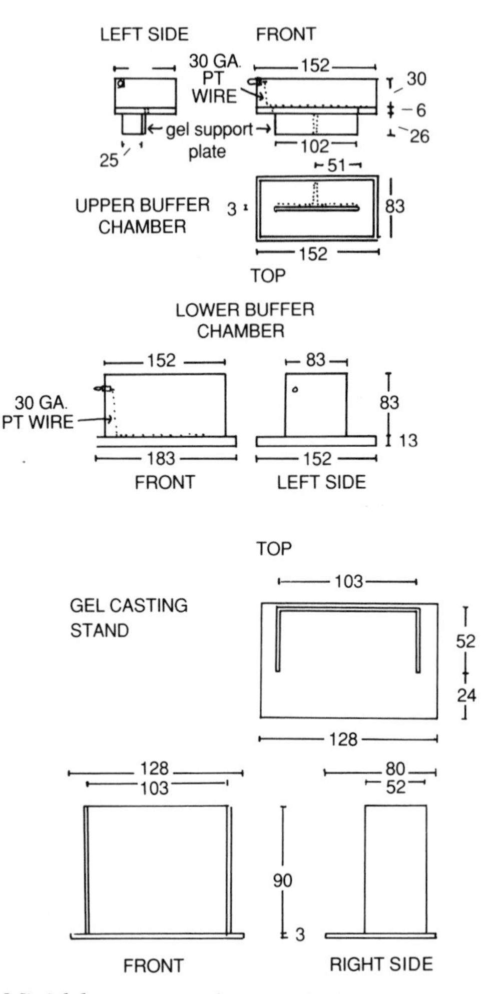

Figure AII.2. Minislab apparatus for protein fractionation. [Matsudaira and Burgess (1978). *Anal. Biochem.* **87,** 387–396.] All dimensions are in millimeters. Plate for front of casting stand, 103 × 90 × 3 mm.

Vertical Electrophoresis (DNA Sequencing)

The plans outlined in Figure AII.3 are for a DNA sequencing gel rig that will run gels about 40 cm long (approximately the same length as standard 14 × 17-inch X-ray film). For optimal utilization of X-ray film, two such gels should be exposed side by side in a standard 14 × 17-inch X-ray cassette. The apparatus is constructed of $\frac{1}{4}$-inch Plexiglas, except for the base. Sets of plates consist of pieces of $\frac{3}{16}$-inch plate glass of the following size: 16 × 7 $\frac{3}{4}$-inch and 17 × 7$\frac{3}{4}$ inch. Spacers are formed from 0.016-inch-thick Delrin, and combs can be cut from the same material, or

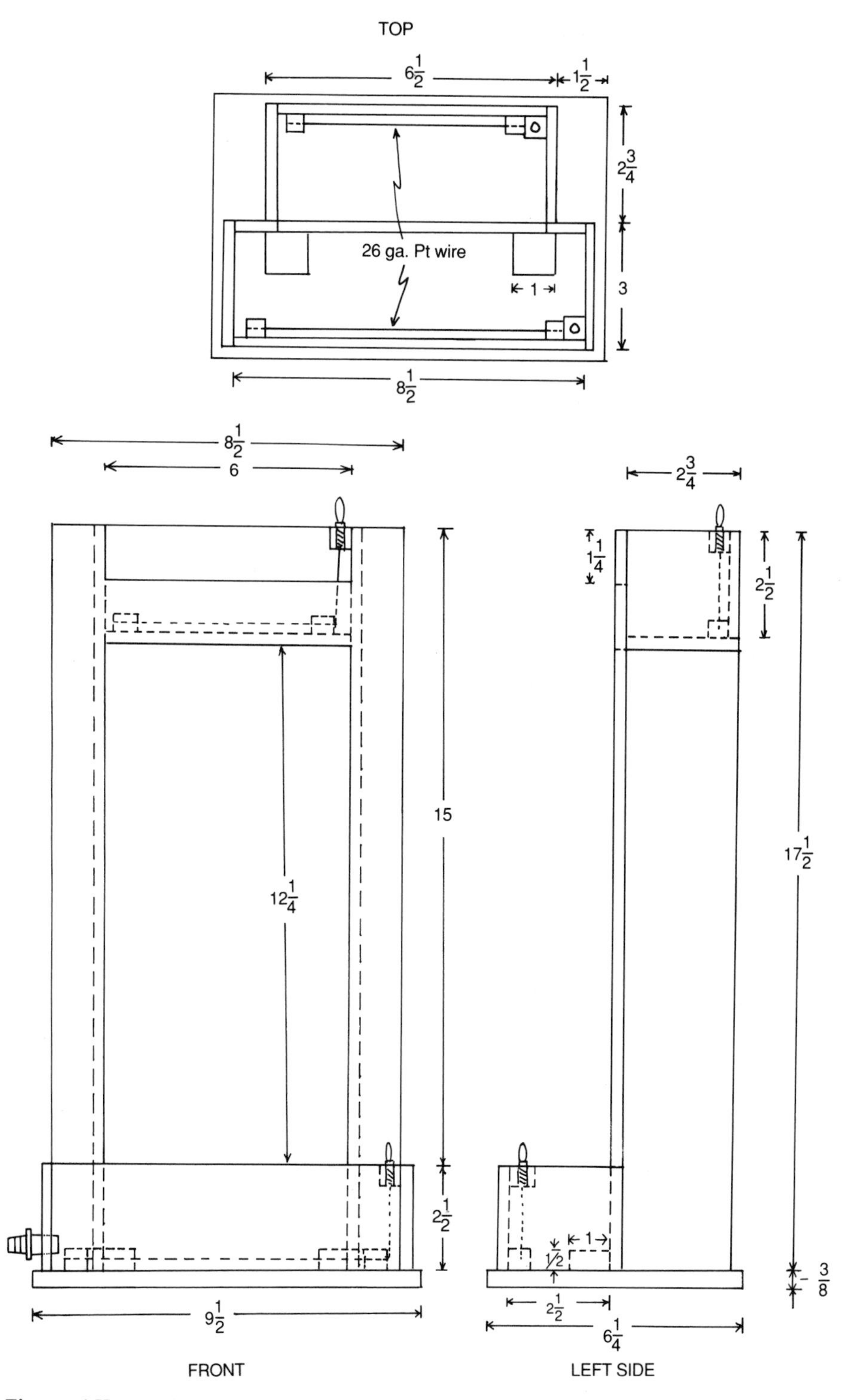

Figure AII.3. A DNA sequencing rig that will run gels about 40 cm long. All dimensions are in inches.

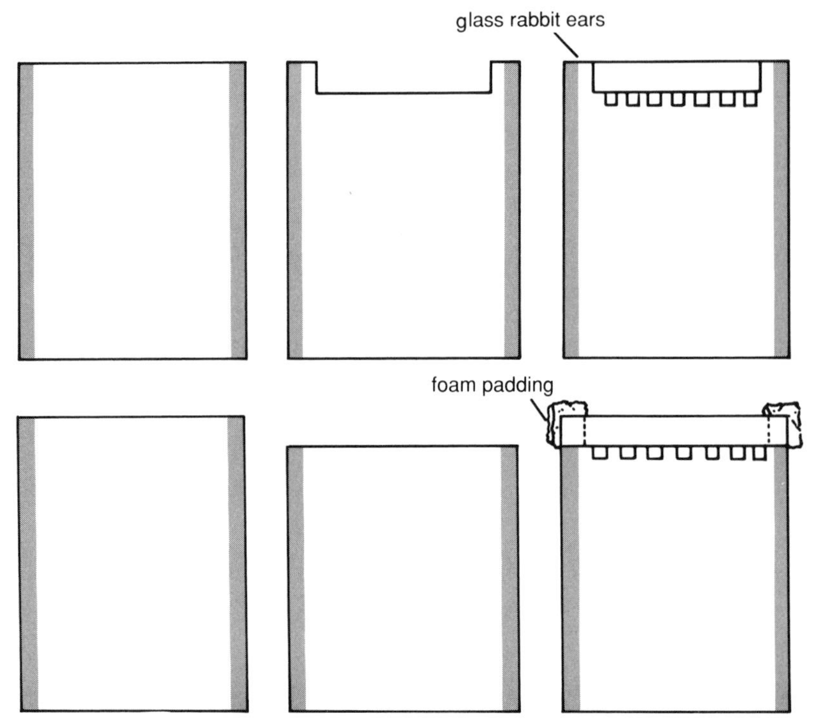

Figure AII.4. Polyacrylamide gel plates, with and without rabbit ears. *left,* Outer plates; *center,* inner plates; *right,* inner and outer plates joined together, showing position of the gel slots.

purchased commercially (Bethesda Research Labs, Inc., makes combs that will fit this gel box design perfectly). In place of the old-fashioned "rabbit ears" at the top of the smaller glass plate, we normally use foam padding (see Figure AII.4). These pads are conveniently made from a foam pad used as a cushion under a sleeping bag (available from L.L. Bean or Eastern Mountain Sports). When the gel is clamped tightly to the box, the pads will compress and form a water-tight seal.

Index